PLUMBING WITH PLASTIC

Other TAB books by the author:

No. 1214
$15.95

PLUMBING WITH PLASTIC

BY S. BLACKWELL DUNCAN

TAB BOOKS Inc.

BLUE RIDGE SUMMIT, PA. 17214

FIRST EDITION

FIRST PRINTING—APRIL 1980

Copyright © 1980 by TAB BOOKS Inc.

Printed in the United States of America

Library of Congress Cataloging in Publication Data
Duncan, S. Blackwell
    Plumbing with plastic.

    Includes index.
    1.   Plastics in plumbing. I. Title.
TH6325.B55   696'.1   79-27712
ISBN 0-8306-1214-9
ISBN 0-8306-9958-9

Contents

Introduction

There was a time, not too many decades ago, when the network of pipes and tubes buried within the framework of our house didn't exist. The outhouse was *de rigueur*, water was drawn by bucket or hand pump, used water was dumped outdoors in any handy spot, and slops were often unceremoniously pitched out of the nearest window. Convenience was nil, sanitary conditions frequently abysmal and health hazards horrendous. A plumber was a man who worked with lead and similar metals—a craftsman or artisan who had never heard of P-traps, shower heads, closet bends or gate valves.

Then came the revolution, and within the space of a few years indoor plumbing became a reality, at least for the rich folks who could afford such luxuries. Indoor running water in a gravity-fed or even a pressurized system became a reality. Waste could be conveniently drained away into a cesspool. The plumber expanded his business into a full-fledged trade, starting with the fabrication of the lead pipes that were so widely used for a time, and expanding into the other areas of plumbing system installation as the years went by and new products became available.

In the early days, installing a complete plumbing system was a difficult chore. Certain skills were needed that most people didn't have and few apparently cared to acquire, and there was a lot of heavy physical labor involved. The plumbing trade was left entirely to those who, for one obscure reason or another, were particularly interested in it. As new materials and products were introduced,

7

new procedures and techniques of installation were developed, and the plumbers expanded their trade even further.

Iron pipe came into vogue, then galvanized steel pipe, and black steel pipe for heating. Brass pipe was introduced as the ultimate in plumbing systems, but never did really work out. For years heavy cast-iron pipe was used for drainage. And all the while, fittings and accessories and fixtures of various sorts were introduced by the hundreds, to cover seemingly every possible plumbing situation.

While the layman paid little attention, the plumbing trade and the plumbing supply industry grew by leaps and bounds, became increasingly complex, and installations got more and more complicated and bound up in rules and regulations. A lot of special equipment and knowledge was needed in order to take care of even relatively simple jobs. As a consequence, the plumbing field has not been one that has gained popular fancy over the years, but instead has remained a rather arcane, "better-leave-it-to-the-professionals" kind of trade. Only a few of the more advanced and adventurous home handymen and do-it-yourselfers took a stab at doing much plumbing work beyond the replacement of an occasional leaking faucet washer.

The advent of copper pipe and tubing, with a full line of matching fittings and associated parts, changed the situation to some extent. This material was (and is) much easier for the home mechanic to work with. No expensive threading equipment is required as with steel pipe, no lead pots and joint leading skills as with cast-iron pipe. Many do-it-yourselfers mastered the art of sweat-soldering and discovered that they could put together a fine and workable copper plumbing system without too much difficulty.

Meanwhile, developmental work continued in other areas of the plumbing field, especially the industrial sectors. The requirements that heavy industry has for piping and plumbing system components are far different than those for residential and light commercial applications, and much more stringent. They need long-lasting and effective equipment capable of withstanding high pressures, handling corrosive or abrasive materials, taking high (or extremely low) heat, and similar exotic characteristics. When the age of plastics slowly came into being, work went forward to develop complete plumbing systems of this incredibly useful and versatile material—or actually, numerous basic materials in various combinations. That development has been eminently successful, and continues today in the quest for more and even better plumbing products made from plastics.

A number of years ago, sufficient components of appropriate design for making complete plumbing installations in homes, commercial buildings and the like were placed on the market. For several reasons the idea didn't catch on, and in fact manufacturers ran into an uphill dogfight. Plumbers didn't like the stuff—because it was too easy to use, too cheap and not enough profit could be made, too new and untried, not widely available and so on. There were plenty of reasons, or psuedo-reasons. Government bureaucrats looked askance at the whole system, decreed that it would never work, would fall apart, wasn't healthful or whatever, and refused in many areas to allow its use at all. Indeed, there were some early problems of various sorts with pipe and fittings, particularly improper installation through lack of knowledge. Probably the system was a bit ahead of its time, not thoroughly debugged, and public acceptance was low.

The situation is different now. Plastic plumbing systems have proven their worth, earlier difficulties have been overcome, and acceptance has become widespread. All types of components are available for residential plastic plumbing systems to take care of just about any installation requirements. Even fixtures are now available in plastic or fiber glass. Plumbers use the products on a daily basis without a second thought. Designers and architects specify plastic plumbing products in their blueprints. Plumbing supply houses stock broad assortments of pipe, fittings, accessories and fixtures for plastic plumbing systems. Today's products conform to the various applicable ASTM (American Society for Testing and Materials) standards, a most important factor. They are approved by the FHA (Federal Housing Administration) and similar agencies on both federal and state levels. Even the plumbing codes, traditionally and historically the last to come around to new products or procedures (even though they are the first hurdle one meets in installing a plumbing system), are now recognizing plastic on a fairly widespread basis as a viable and even superior plumbing material.

Best of all, plastic plumbing products are a do-it-yourselfer's dream come true. Not only is a plastic plumbing system, or some section thereof, one of the easiest large-scale projects a do-it-yourselfer will undertake, it will also afford the best and at the same time the least expensive residential plumbing system that one can have.

And that's what this book is all about—do-it-yourself plumbing with plastic components, from minor repairs and auxiliary systems installation to add-ons in existing systems and complete, full-house installations, including the designing. With a modest amount of pertinent information and a few minutes worth of practical experience you can undertake plastic plumbing installations of whatever sort, in the full expectation of ending up with a quality installation for a modest outlay of cash, time and effort. The key to the whole operation is knowledge, and a complete understanding of just what is involved.

We'll start by breaking down an entire hypothetical residential plumbing system, so that you can discover just how simple it really is—and how it works. Then the various bits and pieces of the system, the nomenclature and the terminology, will be discussed and illustrated—so that you can see what you will be working with and know what parts to request from your plumbing supply house. The necessary tools and equipment will be covered, and so will the rules and regulations governing plumbing installations in general and plastic systems in particular. You'll learn exactly how to work with plastic components and the techniques and procedures used to put things together. You'll be alerted to the possible pitfalls and the potential problem areas, so that you can avoid them. And you'll learn how to design complete or partial systems, install them or troubleshoot them, make repairs, install add-ons or auxiliary systems, and how to connect plastic to metallic systems.

In short, this book will cover the whole field of residential plumbing with plastic components. You will easily be able to relate the information to your own particular plumbing applications or intended projects, to the available supplies and materials in your own area and to any local plumbing codes or other regulations that might pertain.

Never thought you could do your own plumbing? Read on. You'll change your mind, and that's an odds-on bet.

S. Blackwell Duncan

Chapter 1

The Residential Plumbing System

The first step in learning about plumbing with plastic components is an investigation of the residential plumbing system itself. Plugging the parts together is no chore. Anyone who can drive a nail or tighten a bolt can do that. But you have to get the right parts in the right places, and this requires a thorough understanding of a complete plumbing system, what it is, and how and why it works. No two plumbing systems are completely alike as far as specific detail and layout is concerned. All of them, however, function in the same general manner and are composed of various combinations of certain parts. Once you understand how a household system is put together and why it operates as it does, you will be able to design and install such a system to fit any particular requirements.

A logical beginning, then, is to dissect a typical residential plumbing system. Though the discussion will center upon a system such as might be found in a two-story house, with a few modifications the same basic system is applicable to a three-story house, or a single-story house. And though we will be concerned later on with plastic components, the same system can be built using iron, steel, copper or even stainless steel parts. For that matter, the whole system could be made of glass and the general design and operation would remain exactly the same.

A complete plumbing system when seen all at once, or when seen in various peculiar-looking sections roughed into the framework of a house as it is being built, looks mighty complicated and confusing. A small one-dimension drawing of the same system looks even worse. But looks are deceiving. Any plumbing system of this sort can be readily broken down into several subsystems. Some of the systems stand alone and are never interconnected under any circumstances. Others are indeed interrelated and interconnected, and are verbally separated primarily for ease of discussion and clarity of purpose.

Once the system is broken down into the subsystems, either visually or verbally, it is much easier to discuss and understand them. Further, when each subsystem is again broken down into its component parts, the utter simplicity of the whole business is quickly apparent. Each part has an obvious purpose and is in that particular place for an obvious reason. The whole affair is just like a huge Tinkertoy set, and goes together with just about as much difficulty. That is, it does if you have done your studying, design work and preplanning carefully and correctly.

Without water there is no need for a plumbing system in the first place. The whole purpose of the system is to introduce water into a building, and to dispose of used water and certain wastes. The obvious point to begin our dissection of a complete plumbing system, then, is at the source.

WATER SUPPLY

The first stage of a residential plumbing system is designed to extract and convey water from the nearest convenient source into the house. There are a number of different water supply sources and though the object is the same for all, each is handled in a slightly different fashion (Fig. 1-1). Because of this, the plumbing requirements vary somewhat.

Municipal Supply

If you live in a city or a town, even a small town, your water supply is likely to be provided for you. The municipality itself or a water company under its direction obtains raw water from a river, reservoir, deep wells or some similar source, and then processes and purifies the water into potable state in a water treatment plant. The resulting potable water is pumped through a network of pipes throughout a certain designated area and is available to any resident who lives close enough to be able to connect onto the system. In fact, in such situations the resident is generally required perforce to tie into the system and become a customer of the water utility. Billings are made to each water user, generally on a monthly or quarterly basis, to pay for the service. There is usually also an initial

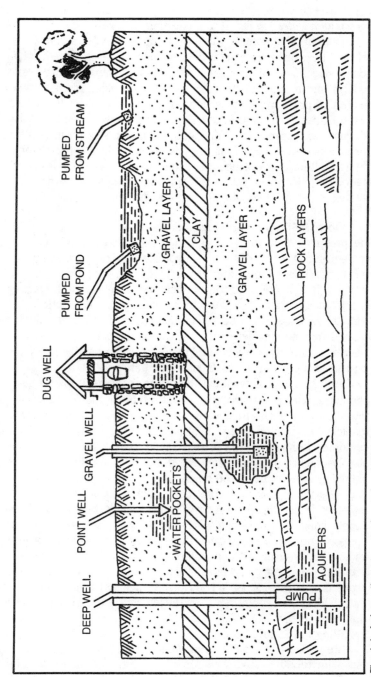

Fig. 1-1. A domestic water supply can be obtained from several sources.

13

charge for connecting to the system, or "tapping in," and this is known as a *water tap fee*. It is a one-time charge.

The pipe that contains the water supply is called a *main*, and will doubtless be located under the pavement of the street in front of your house. This is a pipe of several inches in diameter (sizes vary widely), and in a good system the water will be under a considerable amount of pressure. This is done to insure that all of the water system users will have an ample volume of water and at least adequate water pressure at the house faucets, no matter how much concurrent usage occurs at any given time. That's the theory, anyway, although in practice the results may be somewhat less than marvelous. But most municipal water departments do a fine job, barring major mechanical malfunctions, drought and similar problems.

Once the proper permit has been issued and the tap fee paid, the prospective water user can then go ahead and tap into the system. Exactly how this is done depends upon the particular rules and regulations of the water department involved. One common situation is that the water department itself will install the supply line from the water main to the nearest point just outside, or sometimes just inside, the house foundation. The charge for this service may be included in the tap fee or the water user may be billed separately for the job. Either way, rest assured that the water user pays the bill in one fashion or another.

Another possibility is that the water department may retain a plumbing contractor to make their water taps for them. In this case the contractor and his men will do the job, running the pipe from the municipal main to whatever point near the foundation wall that the customer designates. He will then render a bill for his services, and also be mighty pleased to go ahead and hook up the rest of the plumbing system inside the house.

The third possibility, one that often is found in small communities, is that the water user will have to undertake to install the supply line. This does not necessarily mean, however, that an energetic do-it-yourselfer can actually do the work himself. As often as not the water department will require that the job be done by a licensed plumbing contractor. This, of course, is for their own protection and is probably a good idea. It means that the water user must find his own contractor to do the job, and take care of all details that may pertain.

Installing the Supply Line

Regardless of exactly who does the job, it will be done in approximately the same way. After the position of the main under

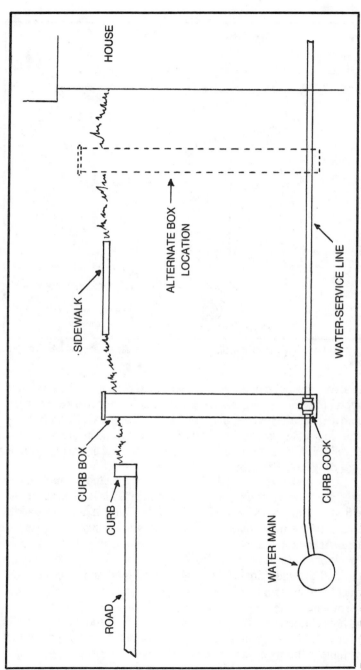

Fig. 1-2. Typical water supply line arrangement showing possible curb cock and box locations.

15

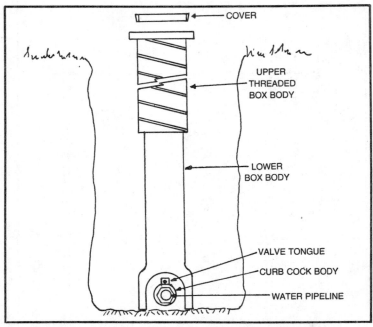

Fig. 1-3. Typical curb box installation protects and allows easy access to curb cock.

the street has been determined, a *street cut* is made and a section of pavement removed. Then a trench wide enough to provide working room and of sufficient depth is dug back to the house. Trench depth varies locale, but is usually at least 3 feet or so, and in any case must be well below local maximum frost-line depth so that the pipe stands in no danger of freezing.

The next step is to attach one end of a suitable supply pipe, called a *water-service line*, to the water main. This is generally done by first bolting a special saddle around the water main. The saddle seals onto the pipe under pressure with a special gasket arrangement. At the top of the saddle there is a tapped hole. Once the hole is positioned correctly, a particular type of valve is threaded into the hole and snugged down. The valve is opened and a special tool attached to it, with a drill bit passing through the open valve body. By rotating the bit a hole is bored through the wall of the main. Once the hole is bored the bit is withdrawn, and even though the main is full of water under great pressure, little or no water escapes. When the bit is withdrawn, the valve is closed and the tool uncoupled, and that's that. This process is called *wet-tapping*.

Lengths of pipe or tubing are then attached to this valve, and run back toward the house to a second valve. The location of this valve depends upon water department requirements. It may be spotted in or slightly to either side of the sidewalk, if one exists, or may be located somewhere on the customer's property, nearer to the house. Usually however, placement is upon municipal property. This *curb valve*, as it is known, is not always required (Fig. 1-2). If the water department does not install one, it's not a bad idea to put one in yourself, somewhere just outside the foundation wall. Then if something goes awry with the water-service line, you will have a means to turn the water off outside the house and thereby minimize potential damage that might occur.

After the water-service line has been continued into the house proper, the curb valve is covered by a special adjustable-length large-diameter pipe, usually made of cast iron with a removable cover on the top. This is called the *stop box* or *curb box* (Fig. 1-3).

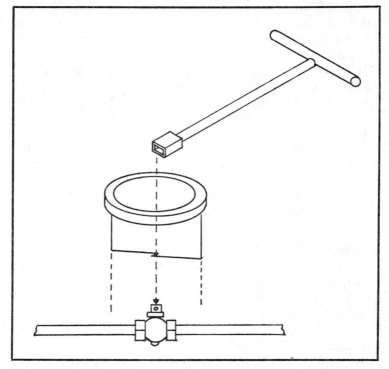

Fig. 1-4. When the curb cock and box is installed without an extension rod, a special valve key in the possession of only authorized persons is used to operate the valve.

The cover protrudes just above ground level. Inside, a long rod extends from the top of the stop box to its attachment point on the valve top. The rod is turned by means of a special wrench, sometimes known as a *key* (Fig. 1-4). By turning the rod counterclockwise through a 90-degree arc the water can be turned on. Turning a quarter-circle in the opposite direction turns the water off very quickly, much more so than an ordinary gate or globe valve.

Variations

Now we come to some variations. As mentioned previously, the curb valve may or may not be required, but probably will be. This may be as far as the water department will go, or it may not be. If no water meter is required, a not unusual circumstance, the homeowner may have to arrange to continue the water-service line into the house. If a water meter is required, the meter may be located just outside or just inside the house, depending upon the locale. The water department may insist upon making this portion of the installation, too. They may supply the meter to the homeowner or require him to purchase and install his own approved meter.

If there is no water meter, the water-service line is run from the curb valve directly into the house and to a *main shutoff valve* just inside the foundation wall (Fig. 1-5). Otherwise, the water-service line runs from the curb valve directly to the water meter. Another variation is that sometimes yet another valve, the same type as is used for a curb valve, is placed in the line just on the street side of the meter, whether inside or outside. This is called a *meter valve*, and serves to conveniently shut the water off if the water meter must be removed for repairs (Fig. 1-6). Again, this portion of the installation may or may not be up to the homeowner. Either way, he'll get to pay for it.

Private Water-Service Lines

There are numerous water supplies that are not municipally operated, but may be privately owned by individual water companies or controlled by resident homeowner groups. Such systems are often supplied by deep wells and may be pressure-fed through distribution mains by pumps, or operated by gravity feed from large storage tanks. These systems are commonly found in exurban and semirural subdivisions and housing developments. Such systems may be very small, serving only a handful of households, or may be as large or larger than many municipal systems.

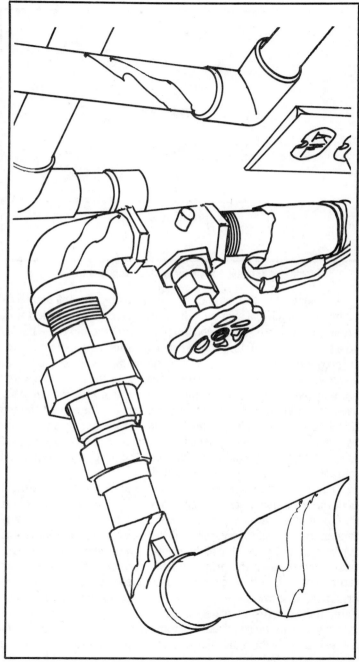

Fig. 1-5. Typical main shutoff valve installation. Here, a galvanized steel supply line riser and brass valve connects to a CPVC system.

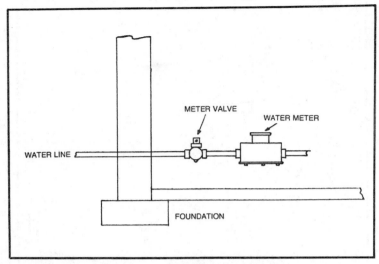

Fig. 1-6. Typical water meter and meter valve arrangement.

Connecting to this type of system is little different than connecting to a municipal system. The water user will need a letter of permission or some similar document allowing the tap to be made, and a tap fee will doubtless be in order (this fee may be included in the property purchase price). The personnel who maintain the water system may be equipped to make taps, or the individual water user may have to find a plumbing contractor to do the job. In many cases it may be permissible for the property owner to do the job himself, provided that he has the expertise, tools and equipment. In any case, it will probably be necessary to hire an excavator to dig the trench. If a new home is being built upon the property, the general contractor can handle the whole affair.

The mechanics of the water-service line installation are just about the same as for attachment to a municipal system. The water mains will most likely be smaller, but are tapped into in the same manner. A stop valve is installed at the main, and a curb valve and stop box placed at the edge of the right-of-way under which the main runs, or perhaps on the property line or in the setback area. The water-service line then continues into the house itself, where it terminates just within the building walls at a main shutoff valve. Water meters are not widely used in these small systems, and probably will not be required. If a meter is required its location will be the same as in the municipal system layout. A meter valve may or may not be included.

Surface Water

Surface water constitutes another possible source that can be easily tapped in some circumstances. This includes still bodies like ponds and lakes, as well as natural springs, creeks and rivers. Over the years this has been a particularly popular method used to supply summer cottages and rural residences.

The installation is made by attaching a *strainer foot valve* to the end of a length of pipe and immersing the valve to a suitable depth in the water body (Fig. 1-7). The water-service line can be laid atop the ground in benign climates, or permanently buried below frost-line in the usual manner. The line runs directly into the building being served and is connected to a water pump. A shutoff valve may or may not be placed immediately on the supply or inlet side of the pump. Using water from this kind of source generally also requires that purification and filtration equipment be installed as well. The water is most unlikely to be potable. Even if it happens to be pure when tests are made, there is always the possibility of rapid and undiscovered contamination at any time.

Point Well

Another source of easily tappable water in some areas can be found just below the ground surface. Sizable water pockets or flow layers can be found in gravel strata lying just a few feet deep. This only occurs in areas where there is a high and fairly constant water table and where the ground structure is such that water permeation is possible on a more or less constant basis.

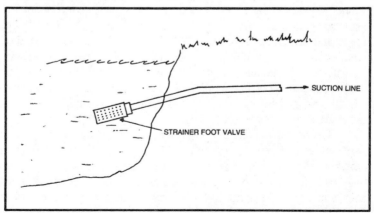

Fig. 1-7. A strainer-foot valve attached to a pump suction line to pick up water from a surface supply source.

21

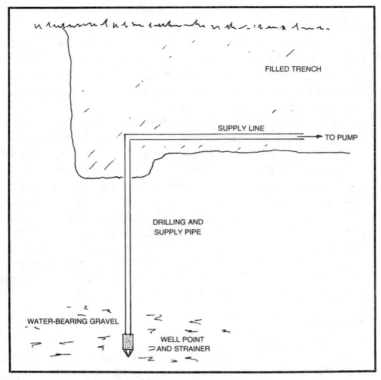

Fig. 1-8. Typical point well installation.

This water is tapped by driving what is known as a *point well*. This is a job that the do-it-yourselfer can easily accomplish, and if conditions are right the labor and materials needed are minimal. A special device called a *well point* is attached to a length of heavy steel pipe and driven directly into the ground like a stake, using tools made for that purpose (Fig. 1-8). Short sections of pipe are added as the point is driven deeper into the ground until sufficient water flow has been achieved. Frequently two or more points are driven in a small area and interconnected to a single supply line in order to provide a reasonable volume.

The water-service line is then extended to the building being served and attached to a pump. Again, shutoff valves may be included or not as desired (Fig. 1-9). The advantage of installing valves is that if the system must be disconnected for repair work, the valves can be turned off and water will remain in the water-service line and/or pump, at least for a good while. This in turn means that when the pump is reconnected, priming probably will not be necessary. There will be fewer difficulties in getting the pump to operate and "take hold."

Gravel Well

A *gravel well* is somewhat similar to a point well, in that it too taps a water supply that is held in a stratum of sand and gravel located just a few feet below ground level. Such wells, though, are usually deeper than point wells, and if properly made can afford a better and more constant supply of water.

A gravel well is made by driving heavy steel pipe directly down into the ground until a water-bearing gravel layer is reached. Water is then forced down the pipe under pressure and pumped back up again. As the water comes back it carries quantities of sand and gravel along with it. This process is repeated until a large cavity is hollowed out at the end of the well pipe. The cavity becomes a reservoir, and after a period of settling out will provide a continuous supply of clear water. The water-service line is connected to the well pipe at the surface, in a well pit, or underground below frost line. The water-service line is then extended back to the building being served, where it is connected to a shutoff valve if desired and a pump (Fig. 1-10).

Dug Well

The *dug well* is another type of shallow well and is perhaps the most familiar one of all. There are literally thousands of dug wells scattered all about the country, many of them 100 years or more old and still providing good water. A dug well is still a perfectly good option for a water source, provided that there is a high water table or some continuous source of water a few feet beneath ground level.

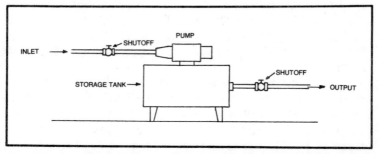

Fig. 1-9. One possible arrangement for shutoff valves in conjunction with a water pump. Unions might also be installed.

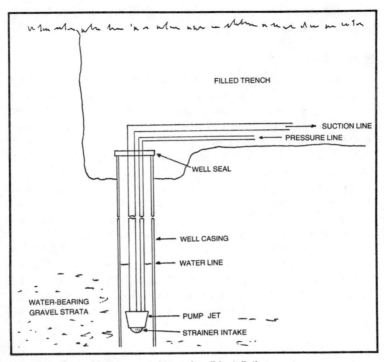

Fig. 1-10. Typical jet pump and gravel well installation.

A dug well is just a hole in the ground, generally from 4 to 6 feet across and anywhere from 10 to 25 or 30 feet deep. They can be larger, and in fact some measure 20 feet across and 50 feet deep or larger. Whatever the size, the hole is dug deep enough so that after a short period of time the bottom several feet will be filled with water constantly. In the old days the walls of such wells were built up of stone laid without mortar, and capped with a small wooden structure to house a windlass or pulley arrangement for a bucket. Today, however, these wells are lined with concrete blocks or large precast poured concrete rings and capped with a cast concrete cover.

In essence the water-service line is installed in a dug well in much the same manner as with the previously discussed water supplies. A strainer foot valve is attached to a length of pipe and suspended in the well a foot or so above the bottom. The pipeline is trenched in below frost line and extends to the building being served, where it connects to a shutoff valve and/or a shallow well pump (Fig. 1-11).

24

Deep Well

When water cannot be found within a distance of 25 to 30 feet below ground level, then it is necessary to go to an entirely different kind of installation. This requires drilling a *deep well* by sinking a borehole down through the underlying strata of soil and rock until an underground aquifer or water pool is intercepted. Such wells are usually at least 40 to 50 feet deep, frequently run between 100 and 150 feet deep, and may upon occasions go down 1000 feet or more. The usual procedure is to drill until a satisfactory flow rate of water is reached, on the order of 3-5 gallons per minute (gpm) minimum for a small single residence, or 5-10 gpm for a more adequate supply.

Deep wells, also called *rock wells* or *drilled wells*, are cased with heavy steel pipe of 4-inch or greater diameter as they are drilled. The water-service line may be installed in any one of several ways. One common procedure is to attach a submersible deep-well pump to the end of a section of pipe and lower it, along with its electrical power lines, almost to the bottom of the borehole. The pipeline is routed out through the casing and laid in a suitable trench below frost depth to the building being served, where it is attached

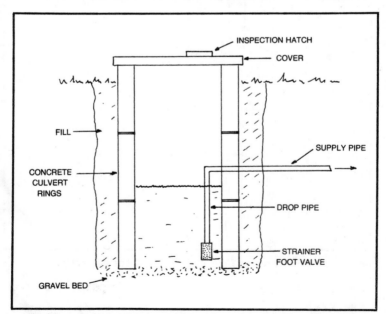

Fig. 1-11. This dug well water supply layout is typical; supply line can be close to surface where there is no danger of freezeup.

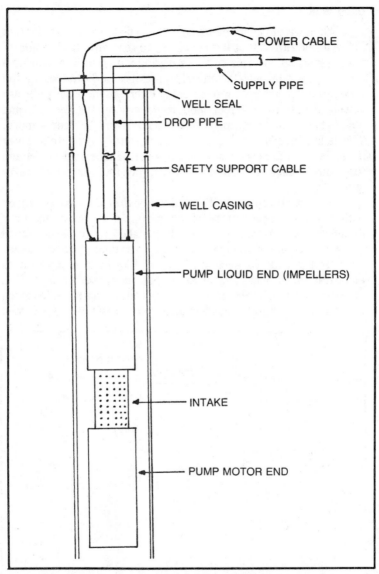

Fig. 1-12. Stylized schematic layout of a typical submersible pump.

to a shutoff valve. The top of the well casing is fitted with a steep cap (Fig. 1-12).

An alternative to this method is to lower a strainer foot valve into the well on a length of pipe and connect it directly to a deep-well pump housed in a small wellhouse built directly over the well. Valves can be placed directly in the line, one immediately on either side of the pump, so that if and when the pump must be removed the water can be cut off. The remainder of the water-service line is then extended in the usual manner from the pump to the building being serviced. Another shutoff valve can be located immediately within the building wall for convenience in servicing or repairing the interior system (Fig. 1-13).

A third alternative is to place a strainer foot valve deep in the well pipe as before, with an unbroken water-service line extending directly into the building being served, attached there to a shutoff valve and/or a deep-well pump (Fig. 1-14).

Any of these methods works just as well as another. The one to be used is generally dictated by job conditions, climate, available equipment, the distance between the well and the building, costs and similar factors.

Gravity Systems

Water delivered through mains by municipal water systems is under pressure when it arrives, and that pressure may be provided

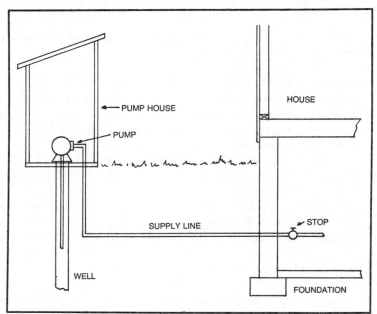

Fig. 1-13. Water pump can rest directly on top of the well casing and in a well house or well pit.

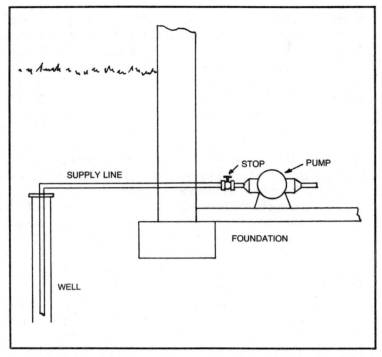

Fig. 1-14. In many installations the water pump is placed in the house basement, remote from the water source.

by pumps or by gravity feed from storage tanks or reservoirs located at a higher level than the mains. Sometimes a combination of the two is used. The result is all the same to the user; when he opens a tap, water pours out. The pressure is always there, at least under normal circumstances.

With the various other water supplies just discussed, water must be drawn or forced into the served building by means of pumps on an individual basis. As long as the pump is running the user has water pressure, and the degree of pressure is adjustable to a certain extent at the pump. If the pump stops, for whatever reason, there is no water pressure and no water flow. Thus, the use of valves in the water-service line is more a matter of convenience than necessity in these installations, to prevent drainback.

But it is also possible to build an individual water supply system serving a single residence without using pumps at all. This is a *gravity-feed* system, and does require the use of valves in the water-service line.

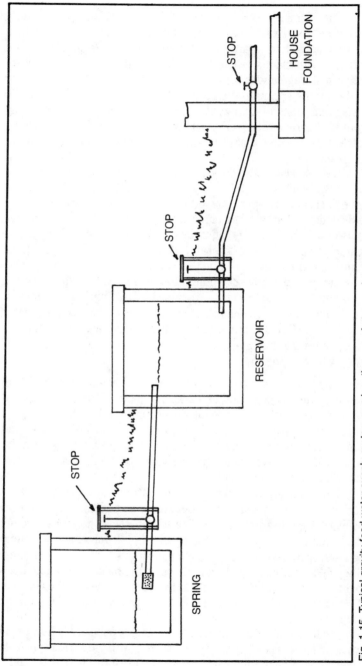

Fig. 1-15. Typical gravity feed water supply system requires the use of stop cocks at each stage.

For example, consider a water supply consisting of a natural spring enclosed with precast concrete rings as a containment, in turn connected to a large nearby storage tank (Fig. 1-15). The spring is continuous, while the tank provides a large surplus of water in storage under constant pressure at the outlet. The supply pipe from the spring to the tank could be fitted with a valve at the spring end to shut off flow here. The water-service line extending from the storage tank to the served building must be fitted with a valve at the tank, so that the water-service line can be shut down at the point of origin.

A second valve should be provided at the far end of the line, just inside the building. At this point, the water in the line is likely to be under considerable pressure depending upon pipe size, tank size and configuration, and the difference in elevation between the two ends of the pipeline. The purpose of this valve is the same as in a municipal pressurized system, since from a user's practical standpoint the effect of the system is the same—water under pressure. In fact, the pressure generated by a gravity system is often considerably higher than that provided by many municipal systems. There must be a convenient and quick method of cutting the water flow in case of emergencies or in order to make repairs or additions to the internal plumbing system.

COLD WATER SYSTEM

The *cold water system* is located essentially entirely within the building, and is comprised of a series of pipes, fittings, valves and assorted gadgetry that extends throughout the building. The purpose of the system is to distribute cold water in an unobtrusive and convenient fashion to various designated usage points where the water is delivered up at the turn of a handle or the click of a solenoid, for whatever purposes are intended.

The starting point of the cold water system is somewhat variable. Generally speaking, it begins at a somewhat indeterminate point where the water-service line terminates as it comes through the building wall. This may be at a main shutoff valve, a water pump or a water meter. In some plumbing systems, especially old ones, the pipeline may come straight through the building wall and keep right on going, without benefit of any fitting, valve or other device. This practice, however, is hardly recommended as being a good one.

So that you can gain some understanding of basically how a cold water system is put together, we will start with the various possible beginnings of a hypothetical system and trace it through to the end.

Water Meters

Water meters appear in a good many designs, shapes and sizes, but they all have one principal purpose in common. They measure the amount of water that passes through the water-service line and is consumed within the building. You probably will not have occasion to install one yourself, and in fact water meters may not even be used in your locale. Many systems do have them, but people often don't understand them. Water meters can be a definite asset to a plumbing system and deserve a bit of investigation.

The design particulars of water meters are of no consequence to us here. What is of interest is how they are read, and how the homeowner can make use of the readings. To begin with, a great many water meters read out in terms of cubic feet of water used. There are various dial layouts, but the one shown in Fig. 1-16 is typical. The fastest-moving indicator is the 1-foot one. The other indicator reads in 10s, 100s, 1000s, 10,000s and 100,000s. By noting the positions of all the indicators and adding the figures shown, you can arrive at a total consumption figure. The difficulty

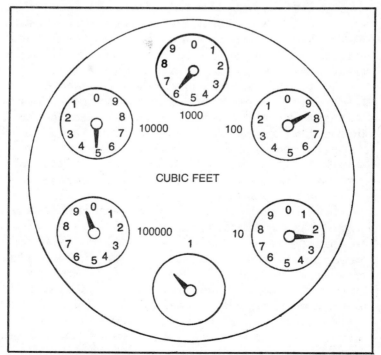

Fig. 1-16. Water meter dial registering in cubic feet.

here is that we don't normally think of water in terms of cubic feet. In order to determine the gallonage, the figure must be multiplied by 7.481—the number of gallons of water in one cubic foot.

The newer water meters that are arranged to read out directly in gallons are much easier to work with. They are also easier to read. The dial face looks almost exactly the same as the odometer in the speedometer of your automobile (Fig. 1-17). Except for the fact that there is no number-drum for tenths, the dial reads just the same as an odometer. All you have to do is read the numbers from left to right to determine the amount of water in gallons that has passed through the meter.

Reasons for Installing Water Meters

Now, even though a water meter may not be required in your situation and it is a rather expensive little gadget, a good case can be made for installing one anyway. Here's why.

First, a water meter is an excellent indicator of hidden leakage somewhere within the system. For example, a quart of water is not very much, but if that amount slowly seeps out of a fitting somewhere within your walls it can create havoc by causing dry rot, soaking insulation and allowing subsequent freezeup, eventually working down into the interior finish, or whatever. Yet the 1-foot, or better yet the 1-gallon, indicator on the water meter will mark in obvious fashion the passage of even so small an amount.

If you suspect that there may be a leak somewhere, all you have to do is make sure that all taps in the building are fully off and all water-using devices shut down. Then watch the meter for a bit. If the indicator shows the passage of water, you have a problem and had best do some hard looking. This is also a handy way to make periodic checks to insure that you don't have any leaks. Shut everything down and watch the meter once every quarter or six months. If nothing moves, you're all set.

Second, taking before-and-after readings on a water meter is a mighty handy way to determine the amount of water you have used for a particular purpose, and the amount you will need to use for the same purpose next time. For instance, you can check to see how much water it takes to fill your swimming pool or garden pond. Or you can fill the pool or pond to a predetermined capacity, according to the meter readings. If you water your tennis court, vegetable garden or lawn on a regular gallonage basis, you can easily determine by reading the meter when you have watered enough for that particular session.

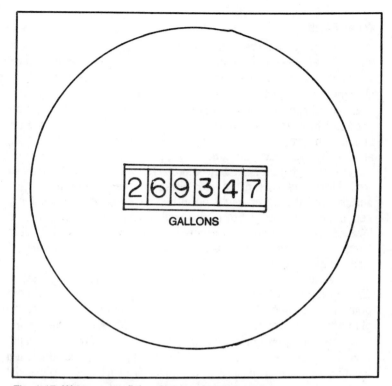

Fig. 1-17. Water meter dial registering directly in gallons.

Third, a water meter can be a great help to those who are conservation-minded and prefer to use as little water as is conveniently possible, or to those who find themselves in a water rationing situation because of drought. By taking periodic meter readings it is easy to determine if an excess amount of water is being used by the household in general. By the same token, it is just as easy to determine when the day's allotment of water has been used up.

Fourth, the water meter reading can be an aid in maintenance programs where pumps are involved. The amount of wear and tear on a water pump and the pump motor is dependent to a large degree upon the amount of water that is pumped. A regular maintenance program can easily be established, based upon cubic footage or gallons of water pumped as indicated by the water meter. A program like this can go a long way towards preventing unforeseen, exasperating and expensive breakdowns in the water supply system, especially in areas where the water is loaded with minerals that can accumulate and cause malfunctions.

Water Pump

The specifics of water pumps and water pump installations are well beyond the scope of this book. However, a few comments are in order. A water pump is unnecessary where the water supply is already pressurized. Where water must be drawn into a building from a local supply source, a pump is necessary. The type of pump used and its location is dictated by the requirements of the particular installation and the equipment available or desirable. Shallow wells are usually equipped with *shallow-well* pumps of either the *piston* or *jet* type, though is some instances *submersible* pumps can be used as well. Deep wells are fitted with either submersible pumps or jet pumps, and the latter might be either *single-stage* or *multistage*, depending upon the depth of the well. The piston or jet pump is located above ground at a convenient spot, while the submersible pump is attached to the end of the water-service line and placed directly in the well.

If a pump is used, it might be in the well, in a pit or well-house directly above the well, within the served building itself, or at some point in between. Regardless of location, the pump must be plumbed into the system. This can be easily done with plastic pipe.

There is another circumstance that might require the installation of a pump of a somewhat different type. This occurs when insufficient water pressure in a municipal system is a perennial problem. Though the volume of water being delivered through the mains could be more or less adequate, the pressure might not be great enough to provide good flow for the fixtures and appliances in the served building.

The answer is to install a *booster* pump at the head end of the cold water system, immediately on the output side of the main shutoff valve in the water-service line and water main. Of course, even this doesn't work where the water volume in the mains is inadequate to serve the customer.

Booster pumps are also sometimes used in an auxiliary capacity in large plumbing systems that include extensive watering or irrigation lines. These auxiliary systems may be separated from and operate independently of the cold water system.

Storage Tanks

Many water supply systems that include a pump also are equipped with a *storage tank*. These tanks range in capacity from around 12 gallons to as much as 85 gallons, and are an essential piece of equipment for proper pump and water supply operation.

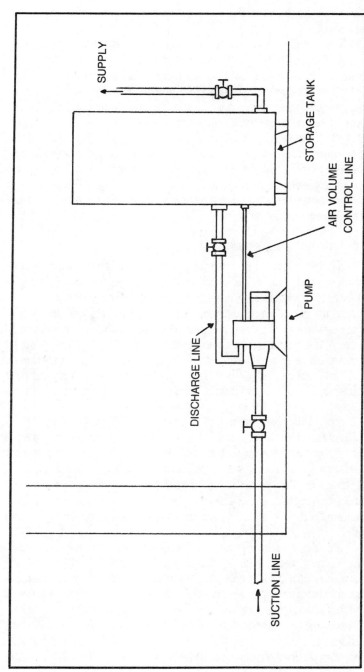

Fig. 1-18. Typical water pump and storage tank plumbing.

The only exception is the *constant-pressure* pump which uses no tank at all.

The smaller tanks are often made so that the pump can be mounted directly on an integral mounting pad or plate. The larger tanks are freestanding and can be located at any convenient spot. Except in the case of submersible pumps, this location is usually right next to the pump itself. There are many kinds and configurations of tanks, some using air-volume controls and some not, but they all are plumbed directly into the discharge or output side of the water pump. Valves are fitted at both input and output of the storage tank (whose contents is always under pressure) to facilitate service and repair work on other parts of the system (Fig. 1-18). Where a booster pump installation is necessary, it too will require a storage tank, often a fairly large one.

Pressure-Reducing Valve

While low pressure is certainly annoying, high pressure is even more so and can cause considerable damage to boot. There is simply no advantage to having any more water pressure than is necessary to satisfactorily operate the fixtures and appliances on a given system. The adequate pressure rating varies from system to system, but in general for a residential system 30 pounds per square inch would be considered low, 45-50 about right in most instances and 70 would be high. When the pressure gets over 80 pounds per square inch, various difficulties are likely to be experienced with the system, or at least that potential exists.

For most residential applications 45-50 pounds is sufficient, a bit less for single-story and relatively small houses, a bit more for three-story and/or large houses. But it frequently happens that pressure in a municipal water main is much higher than this, perhaps as much as 200 pounds. Obviously this can mean trouble in a direct-connected residential system. The answer to the problem is to install a device called a *pressure-reducing* or *pressure-regulating* valve.

Pressure-reducing valves are made to fit specific pipe sizes and are factory preset to a particular pressure rating; 45 pounds is a common setting for small-building systems. But they are also adjustable through a certain pressure range, such as 40-80 pounds. If the factory setting is too high or too low for a particular installation, it can easily be changed with a screwdriver. The valve automatically reduces the incoming water pressure to whatever level (within its range) is desired. At the same time, if there are fluctuations in water

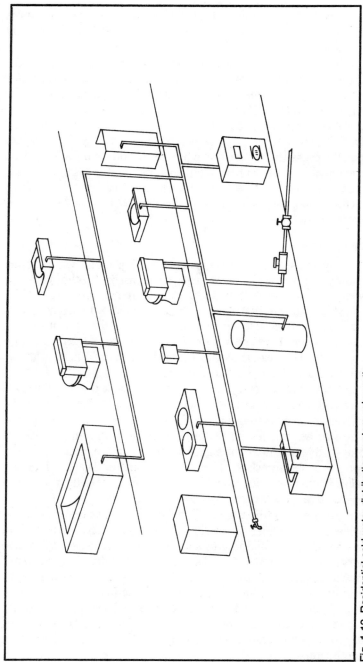

Fig. 1-19. Residential cold water distribution system schematic.

pressure in the main, the valve will smooth out any ups and downs that occur above its set pressure level and deliver a constant pressure.

The pressure-reducing valve is often located at the end of the water-service line, on the output side of a main stop valve, water meter or at some other convenient location at the head end of the cold water system. There are also occasions when it is desirable to locate the pressure-reducing valve close to the connection at the main. This would be the case where the main pressure is higher than the pressure rating of the pipe used for the water-service line. Polyethylene pipe with a 100-pound rating, for instance, could hardly be expected to withstand 175-pound main pressure. So the pressure-reducing valve must be installed at the head end of the water-service line.

Supply Lines

Once past the initial equipment—main shutoff, water meter, water pump, storage tank, reducing valve, assorted other valves or whatever—the cold water distribution system proper begins (Fig. 1-19). The first portion of the pipeline is generally the same size as the water-service line, except in the case of pump installations (pump suction lines are considerably larger than their discharge lines). The pipe size is usually nominal 1-inch or ¾-inch, with the latter perhaps being the more common.

This pipeline continues full-size throughout the building, passing by the points of greatest concentrated cold water usage until it arrives at the next-to-last point. Here the line is often reduced one pipe-size and carries on to the final usage point. In practice, of course, this is seldom a straight-line run, but more often consists of two or more branches. These large-diameter supply pipes are called *risers* when they travel vertically from floor to floor, and may be called *laterals* when they travel horizontally.

As the supply lines pass by various areas of water usage in the building, *branch* lines of smaller size are tapped off the line to feed individual fixtures and appliances. These drops may be employed to serve individual items like a dishwasher or laundry sink cold-water faucet, but may also serve two or more low-demand fixtures like a pair of adjacent lavatory basin taps. Sometimes a reduced size branch pipe may be carried to a usage area, reduced again in size and carried forward to another usage point. The specific pipe sizing depends upon a number of factors, and will be investigated in further depth later on.

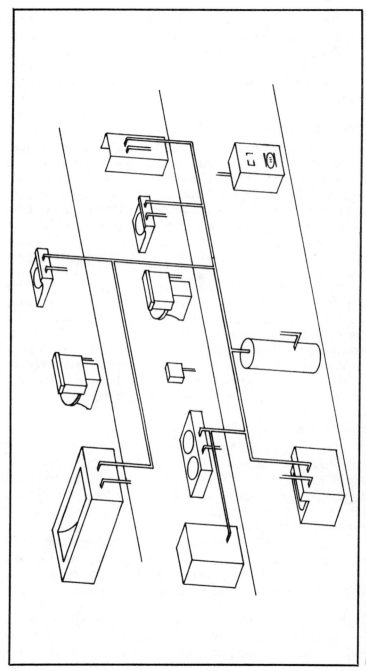

Fig. 1-20. Residential domestic hot water distribution system schematic.

39

The end result, however, is a spiderweb of pipes of varying sizes that eventually feed every single cold water utilization point in the building. Complicated though it may look, the entire system is comprised of a simple composition of pipes and fittings, with an occasional valve thrown in and a series of adapters as necessary to connect to fixtures and appliances.

DOMESTIC HOT WATER SYSTEM

We Americans for the most part have long since gotten past the drudgery of heating our water atop a stove or over a fire, even though a good share of the remainder of the world has not. We would feel much put upon if we couldn't draw piping hot water from the tap with a flick of the wrist. In fact, we tend to become irate even when the shower stream commences to turn tepid. So as far as we are concerned, a faultlessly functioning system of providing limitless hot water is second in importance only to having running water in the first place.

The opposite number of the cold water system is called the *domestic hot water system* (Fig. 1-20). The term "domestic" is used to indicate that this system is separate from a heating or any other kind of system that uses hot water as a medium. In current construction practices, hot water used for heating never comes into contact with that used for domestic purposes; there is never any interconnection whatsoever. The domestic hot water system is smaller and even simpler than the cold water system.

Water Heaters

The starting point for a domestic hot water system is with a *water heater*. The water heating unit, which may be located virtually anywhere in the building, is connected at its inlet to the cold water supply. The water within the tank is heated by flame, electricity or heat exchange, and flows through the outlet and into the domestic hot water system proper.

The most common installation is the *tank-type* unit, which makes use of a special water tank ranging in capacity from 10 gallons to 82 gallons in any one of numerous designs and configurations (Fig. 1-21). These are freestanding units that can variously be placed in utility rooms, basements, small closets or even under kitchen counters. The heat source may be provided by electric heating elements or burners fired by natural gas or propane. The tank is most often located directly adjacent to the point of greatest hot water usage, or as equidistant between two or more points as is

40

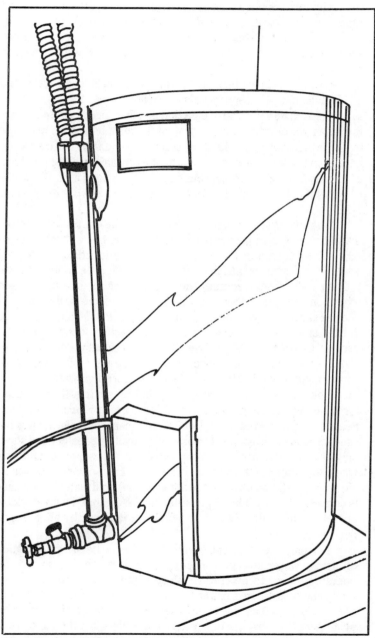

Fig. 1-21. Typical small tank-type water heating unit.

conveniently possible, in the interests of economy. If a large central supply of hot water is required, two tanks can be plumbed in tandem. Often, though, it is more practical to install two or more separate small tanks, each with its own small hot water distribution system.

Another method of supplying hot water can be used where the comfort heating system includes a boiler. Here the cold water supply is attached to a coil of pipe that is either directly immersed in the boiler water or is wrapped around the main hot-water lead pipe from the boiler. As the cold water travels through this coil it picks up heat from the hot boiler water, without ever making direct contact, and finally flows into the distribution system as hot water. This type of water heater is called a *tankless* or *instantaneous* heater (Fig. 1-22).

Whatever the kind of heater employed, the domestic hot water distribution system begins at the outlet of the hot water heater or tank. The makeup of the system is virtually identical to the cold water supply system, and in fact the two parallel one another a good portion of the time. Generally the two lines are run side by side and where one goes the other does, too, using the same kinds of fittings and the same approximate lengths of pipe sections. Valves are used in the same way and adapters are employed to connect onto the various fixtures and appliances. Only occasionally does the cold water system branch off alone, to supply a toilet tank, hose bib or some other usage point that does not require hot water.

There is one water heater that eliminates altogether the need for a hot water distribution system. This is the *instant tankless* heater, a type just recently placed on the market in this country but used for years in England. This is a very small unit designed to serve only one utilization point, and it operates automatically and only when the water is turned on. The units are powered by gas or electricity, and can be mounted anywhere. A cold water supply pipe is connected to the inlet. The outlet can be equipped with a flexible or "telephone" shower head, or plumbed to a tap with a short length of pipe or tubing.

Not only does this kind of unit eliminate the need for a good deal of piping, water storage capacity and tank space, it also reduces operating costs to practically nil by comparison with a standard tank-type heater. In a plumbing system equipped with such units one could say that there is essentially no domestic hot water system, only a series of hot-water supply units attached to the cold water system.

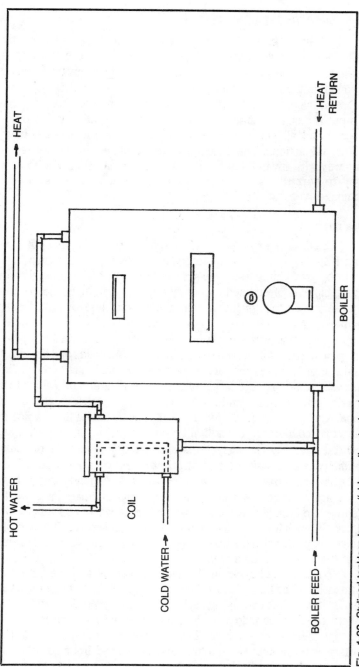

Fig. 1-22. Stylized tankless-type or "sidearm" water heater.

HEAT

HEAT RETURN

BOILER

HOT WATER

COIL

COLD WATER

BOILER FEED

43

THE DRAIN-WASTE-VENT SYSTEM

The drainage system of the house, which is also frequently called the *drain-waste-vent* system, appears to be the one that is least understood and most often incorrectly or haphazardly installed by amateur plumbers. This system is in actuality comprised of three subsystems, one for drainage, one for waste and one for venting. To put it a slightly different way, though the system is interconnected and all one piece, certain sections of the system have different functions. The drainage system is never interconnected in any way with any of the other plumbing systems. About the easiest way to sort this system out is to discuss it in three sections: drainage, venting and sewage.

Drainage

The drainage system is comprised of a series of pipes, fittings and traps—no valves—that carry used water and human waste away by gravity into a sewage system (Fig. 1-23). The general term "drainpipe" is often used to signify any one pipe or section of the system, even though the various parts have their own particular names.

A pipe that carries nonhuman liquid waste is called a *waste* pipe. If a pipe carries human wastes made up of both solids and liquids, it is a *soil* pipe. Pipes that are installed horizontally are often called *laterals*, while the vertical pipes are *stacks*. A lateral or a stack that carries human waste and nonhuman waste together is also a soil pipe. Often two or more fixtures will drain into one line, which in turn may connect to yet another line and eventually go into a larger, central line. These subsystems are called *branches*. A principal branch of a stack may be called a *primary branch*, while others may be *horizontal branches* or *secondary branches*, depending upon how they are arranged. A stack that receives nonhuman liquid waste (sometimes called *gray water*) and/or human waste as well is a *soil stack*. If a stack receives only nonhuman liquid waste, it is a *waste stack*. A principal pipe that receives waste from several smaller ones is often termed a *main*.

To illustrate how a simple system might be arranged, we can start with a one bath plus kitchen sink setup. When you dump used water in a sink it travels through a basket-type strainer and into a short tailpiece or tailpipe. The same is true of a lavatory basin, except there is no strainer. This short length of pipe carries the waste water directly into the trap and on into the drainage line. The section of drainage line between the trap and the nearest other drain

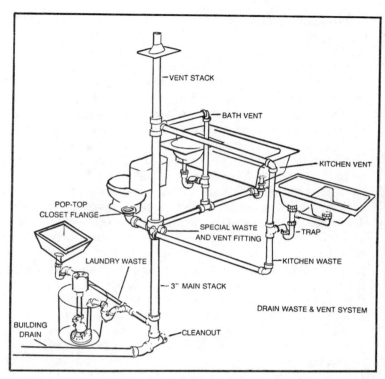

Fig. 1-23. Typical compact residential drain-waste-vent (DWV) system (courtesy of Genova, Inc.).

pipe is called a *fixture drain*. The fixture drain may also be designated by its particular function, i.e., a *sink drain* or a *lavatory drain*.

This pipe connects to a larger one, which in a simple system is likely to be the soil stack. If these particular fixtures happen to be a considerable distance from the soil stack, they may be connected to a branch line or to a waste stack. If the connection is made through a branch line, that branch will eventually connect with either a soil stack or a waste stack. Either or both of these stacks then drop down to a horizontal line at the lowest point of the drainage system, a large-diameter pipe known as a *house drain* or *building drain*. This pipe is the ultimate collector of all drainage from the system. It exits the foundation at some convenient point and terminates approximately 5 feet from the building. From that point on the sewage system takes over.

The tub or shower in our simple system is connected in much the same way. Used water flows through a drain, into a short length of pipe and to a trap, which is located as close as is conveniently possible to the fixture. The water then flows through the fixture drain and into the soil stack or other larger drainage pipe.

A toilet is installed a bit differently. Since it is already equipped with a built-in trap, a second one is not used. A floor-mounted toilet is affixed to a special fitting called a *closet flange* or *floor flange*, which in turn is connected to a *closet bend* and a short length of soil pipe. The soil pipe runs directly into the soil stack. A wall-mounted toilet drains through the rear and must be mounted directly to a special fitting that is installed in the soil stack. In both cases, the sewage drops directly down the soil stack and into the house drain.

Although there are a great many variations as far as specific hookups are concerned, the underlying principles remain the same. There are certain guidelines which must be followed to make a proper installation; these will be explained in greater detail in later chapters. One of the most common variations takes place when two or sometimes three adjacent fixtures—a double sink and a laundry tub, for instance—are connected to a common trap and fixture drain. This is known as a *continuous waste pipe*. Each individual fixture could have its own trap and fixture drain, but a continuous waste pipe works just as effectively, if properly done, and saves material and labor. Sometimes such fixtures, like a pair of lavatory basins, are fitted with individual traps which lead to a single fixture drain.

Traps

One of the most important parts of any drainage line is the *trap* (Fig. 1-24). It is absolutely essential that every single fixture on the system—tub, shower, sink, wash basin, washing machine, dishwasher or whatever—drains directly into a trap. Most fixtures are connected to individual traps and every toilet has its own built-in trap. In some cases, however, it is permissible to attach two or even three fixtures to a single trap.

Traps are located as close as is conveniently possible to the fixture. In most cases, this point is at the end of a very short length of pipe that carries drain water from the fixture proper. There are many different configurations of traps that are used to fulfill various specific installation requirements, such as the *running trap, drum trap, P-trap, S-trap* and so on. But they all serve the same purpose. The trap is always filled with water, and the water serves to seal the drainage system off from the building's interior atmosphere.

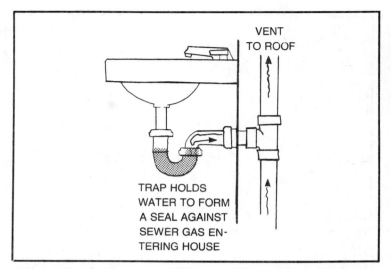

VENT
TO ROOF

TRAP HOLDS
WATER TO FORM
A SEAL AGAINST
SEWER GAS EN-
TERING HOUSE

Fig. 1-24. The principal function of a trap is to prevent sewer gas entry into the house, as well as aiding in proper drainage from the fixture.

The reason for this is simple enough; it's a matter of health and safety. Sewer systems are not only loaded with all manner of lethal bacteria; they are also charged with an interesting variety of gases under low pressure. These gases can be extremely poisonous, highly explosive or very corrosive, and often all three at once. On top of that, the odor that emanates from a sewer (its least danger-ous aspect) is most obnoxious and nauseating.

Since there is no way to overcome these characteristics of a sewage system, the gases must be prevented at all costs from entering the building. The small plug of water located in each trap serves that purpose. And it does so with great effectiveness and in an ingenious fashion. Because water always seeks its own level, the water quantity in the trap is constant. It cannot drain away under normal circumstances and it does not stagnate because it is con-stantly replaced as the drain is used. Human error is avoided since there are no valves to be opened and closed. The trap never needs to be checked to make sure it is full, and, in short, no heed need be paid to it at all except for an occasional cleaning out if it should become clogged.

Cleanouts

Another important part of the drainage system is the *cleanout* (Fig. 1-25). This is a special fitting that is installed to facilitate

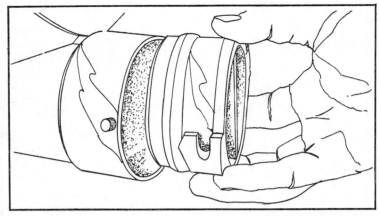

Fig. 1-25. Typical DWV system cleanout opening and twist-to-lock cleanout plug.

cleaning the drainage lines should they become plugged for any reason. Cleanouts, or *cleanout plugs* as they are commonly known, are found on the better grades of one-piece traps, so that unclogging a fixture drain becomes a simple matter of unscrewing the plug and draining the trap. Cleanouts are not normally used on any of the smaller portions of the drainage system, since those lines can be augered out easily enough by entering through the fixture drains themselves.

However, cleanouts are nearly always installed where a soil stack or waste stack joins the building drain. In extensive drainage systems cleanout plugs might also be placed at the ends of long branch lines where they connect to a stack. A simple system, then, would only have one cleanout plug, located where the stack takes a 90-degree turn to enter the house drain. By and large, you can figure that wherever a stack makes a direction change of 45 degrees or more, or where there is a pipeline distance of more than 45 feet from the last cleanout, a cleanout plug should be installed.

Venting System

Although we have chosen to separate the venting system from the drainage system as a matter of convenience in discussing it, the two are actually interconnected and frequently use the same pipes. The venting system is essential in allowing the drainage system to function properly. Without venting, drainage would be a virtual impossibility and since the traps would not work properly either, the health hazard would be extremely high.

The venting system serves several purposes. First, the presence of an unrestricted air flow through the vent pipes allows waste liquids to drain away quickly and efficiently through the drainage pipes. If the drainage pipe system were completely closed, the flow would likely be slow and sluggish and solids would not be carried along in the stream. This would be like trying to pour liquid from a can whose only opening is at the spout; the flow is slow. If a vent is opened in the can top, the flow is rapid.

Second, this free air supply prevents water from being siphoned from the various traps. A strong flow of water through a pipe creates a vacuum behind the water body, at the upper end of the pipe. The suction thereby created could easily empty a trap, pulling the trap water right down the drainpipe. With proper venting this cannot happen.

Since sewer gases are always present in the sewage lines, they can easily travel back along the pipes and up into the drainage system. In a properly vented system, any gases that do accumulate keep right on going and are exhausted into the outdoors where they

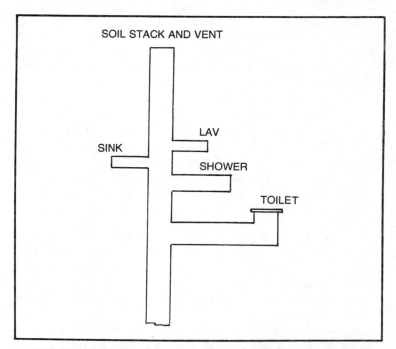

Fig. 1-26. Typical direct venting arrangement where fixtures are vented by the stack into which they drain.

can do no harm. Were the vents not present, sewage gases could conceivably build up enough pressure to bubble through the water seal in the traps and into the relatively closed atmosphere inside the building.

And last, a constant circulation of fresh air through the drainage system insures that the growth of slime and algae is kept to a minimum. Also, gases or condensates that might cause corrosion to the piping (this is not a factor with plastic plumbing) are minimized.

There are a number of rules and regulations governing the placement and installation of vents, and these details will be investigated a bit later. As you might expect, there is also some terminology involved. In the simplest case, provided that the fixtures are close enough together, they can be drained and vented in the same pipe (Fig. 1-26). This is called *direct venting*. It also is a form of *wet-venting*, which occurs when any portion of a waste pipe serves as a vent for other fixtures attached to the same pipe. This situation occurs, in fact, with all fixtures even if only for a very short distance.

In more complex drainage systems, however, the fixtures are often scattered about the premises at considerable distances from the soil stack. In such situations the fixtures cannot be properly drained and vented simultaneously through the soil stack, and perhaps not through an additional waste stack. Then the venting system must be enlarged and more pipes put in. There are a number of methods of doing this.

Back venting is commonly used, where a pipe is connected close to the fixture drain and runs up and back to the nearest soil or waste stack (Fig. 1-27). Two or more fixtures can be back vented by a system called *circuit venting* (Fig. 1-28). A fixture can be individually vented with a separate pipe extending directly through the roof to the outdoors (Fig. 1-29). Sometimes two or more fixtures are piped to a common vent pipe. The common vent pipe might then connect to stack, or might continue directly to the outdoors.

A vent pipe that rises directly through the roof is called a *vent stack*. In some cases it might also be necessary to include *relief vents*, which are pipes running from one waste or soil stack to another in order to afford air circulation between the two. *Branch vent* is the term given to a section of vent piping that picks up two or more fixtures and eventually terminates at a stack.

The main object of a venting system, however it is specifically piped, is to maintain good air circulation throughout the upper portion of the drainage system when it is draining, and through

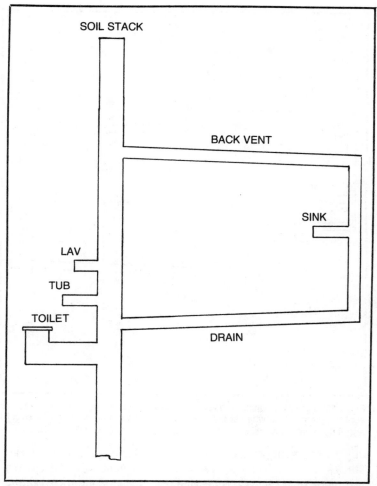

Fig. 1-27. Typical back venting arrangement. Fixtures at the left drain into and are vented by the stack, while sink at right drains into stack but is vented by separate line going back to same stack.

almost the entire system when it is not so as to maintain equal air pressure throughout the system. Each fixture trap must be adequately vented in one fashion or another, with the vent being of sufficient diameter and in correct proximity to allow proper functioning.

Erring is most often done in the direction of too few vents, and the presence of too many vents is an unlikely situation. Nowadays, the practice of *double-venting* is a common one, whereby there are

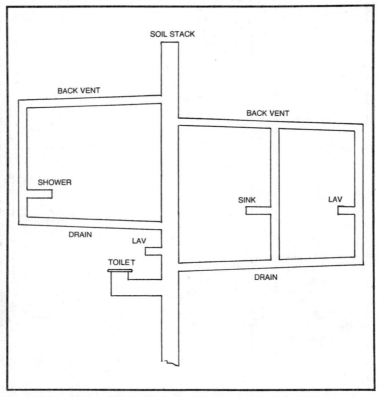

Fig. 1-28. Typical circuit or loop venting. Fixtures at lower left drain into and are vented by the stack, while shower at left has separate back vent (one circuit) and fixtures at right are vented by double back vent (second circuit).

at least two vent pipes exiting the roof. This might consist of a soil stack and a waste stack, or a soil stack and a vent stack. Often there are more than two. Considerable use is also made of *re-venting*, which consists of using any of the methods of venting previously discussed (and others as well) except for direct venting.

There is another type of vent sometimes used in drainage systems, too. This is the *fresh-air inlet* or *fresh-air vent*. It isn't required by the National Plumbing Code; nor is it generally required by local codes. In some areas, however, this vent is installed as a matter of course and a few local codes do insist upon it. The fresh-air vent is generally installed at a handy point in the house drain just before it exits the building. If a house trap is installed, the fresh-air vent is put in just on the house side of the trap. Unlike other vents, this one does not continue upward to exit through the roof. Instead, the pipe rises a bit and turns through the wall, terminating in a screened cap just above ground level (Fig. 1-30).

The fresh-air vent has many good points. The admittance of fresh air at nearly the lowest point in the drainage system allows excellent circulation throughout. Also, the open vent at such a low point in the system acts as a safety valve should blockage occur in the sewage line. Backed-up waste will flow out of the vent rather than gurgling up through various fixture drains within the building. The blockage problem becomes immediately noticeable without causing any particular damage. Blockages that might occur on the house side of the fresh-air vent, of course, still remain problematical. About the only drawback to installing a fresh-air vent is that the job requires a little extra work, and a few additional parts that may

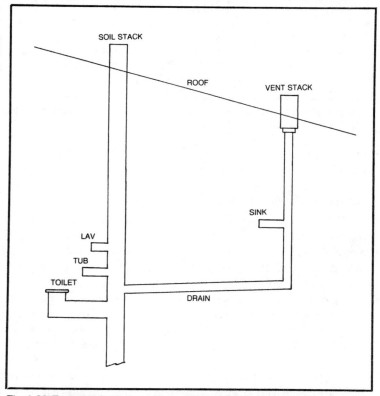

Fig. 1-29. Typical individual vent arrangement. Fixtures at the left drain and vent through stack, while fixture at right is too far from stack for proper venting there. Instead, it is fitted with an individual vent stack to the outside. Note enlarged top at point of roof exit, used in cold-weather areas to prevent frosting over that can occur with a small stack.

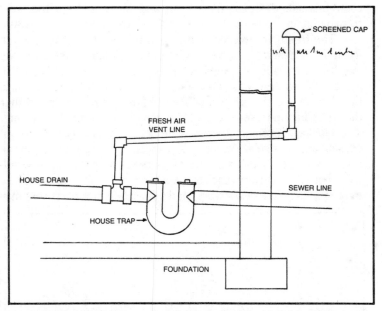

Fig. 1-30. Fresh air venting system allows plenty of air circulation and maintains constant air pressure in DWV system.

actually not be required by code. But all in all the installation makes good sense for any system, required or not.

WASTE DISPOSAL SYSTEM

The last portion of the waste system is concerned with emptying the house drain of all waste materials and conveying them away into a *sewage system*. In one way or another, the waste, now called sewage, is permanently disposed of. Not many decades ago sewage, known then as "night soil" or more crudely, "slops," was dumped into yards and streets or drained off into cesspools and open ponds. Today's regulations are just a bit more stringent. You have two basic choices. One is to tie into a public sewage disposal system. The other is to provide yourself with an individual septic system.

Much to the consternation of many unknowledgeable owners of new unimproved property, there is also a third alternative—no system at all. Many areas disallow the installation of a septic system if soil conditions are not suitable for rapid water absorption. And since privies are now generally outlawed, the proposed new house can't be built.

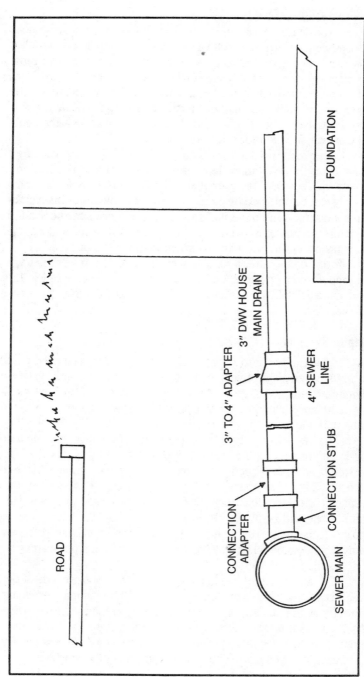

Fig. 1-31. Typical layout, simplified and stylized, for a municipal sewer main hookup.

ROAD

3" TO 4" ADAPTER

3" DWV HOUSE
MAIN DRAIN

CONNECTION
ADAPTER

4" SEWER
LINE

CONNECTION STUB

SEWER MAIN

FOUNDATION

55

Public Sewage Systems

If you live in a city or a community that has its own public sewage disposal system, you will be required to tie into it whether you want to or not. In fact, such systems are by far the most satisfactory for household waste disposal, especially since problems are very seldom encountered and the homeowner need not be concerned about waste disposal at all. The stuff goes down the pipe and disappears, and that's that. This is also generally the most economical situation for the homeowner.

Connection to a municipal system will most likely be made by a crew from the sanitation department that operates the system. There will doubtless be a permit and a sewer tap fee involved, which usually can be obtained by either the property owner or his plumbing contractor. The municipal sewer main will probably be located beneath the street in front of the property, or perhaps along the back property line or in an easement held for the purpose.

In any event, after a suitable trench is dug, the sewer main will be tapped and a sewer pipe extended onto the property to join the end of the building drain (Fig. 1-31). The trench is filled in, and that's all there is to it.

Septic Systems

If you can't tie into a municipal sewage system, you must build a tiny sewage plant of your own. This is known as a *septic system*. Though there are a great many specific variations in how such systems are built, they consist basically of two elements, a large tank and a leaching or absorption field.

The system layout begins at the termination point of the house drain, where a length of sewer pipe is attached (Fig. 1-32). The sewer pipe leads directly to a septic tank which is buried in the ground. Tank sizes vary, as do their shapes, but nowadays the capacity is seldom less than 1000 gallons for residential use. The tank should not be placed closer than 5 feet from the house, and is often placed about 10 feet away. But local conditions and/or codes may dictate some other tank location, so the sewer pipe run may be fairly long.

The sewer outlet line is connected to the opposite side of the septic tank and runs to the *absorption field*. This pipe may terminate directly at the absorption field site, or may run into a *distribution box*. This is a small tank to which several *disposal* pipes are connected, so that the effluent discharge is routed more or less equally into each disposal pipe. The disposal pipe system, which also may

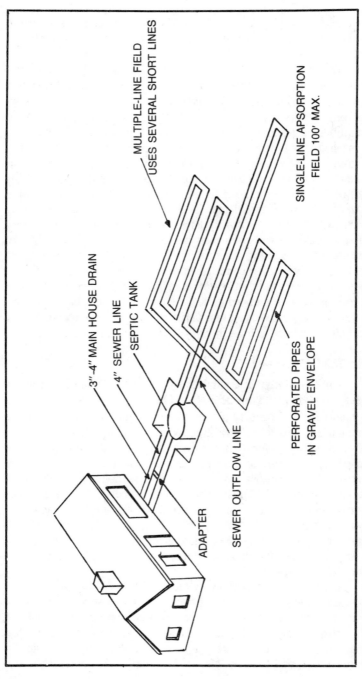

MULTIPLE-LINE FIELD
USES SEVERAL SHORT LINES

SINGLE-LINE APSORPTION
FIELD 100' MAX.

3"–4" MAIN HOUSE DRAIN

4" SEWER LINE

SEPTIC TANK

ADAPTER

SEWER OUTFLOW LINE

PERFORATED PIPES
IN GRAVEL ENVELOPE

Fig. 1-32. This diagram shows the principal parts and arrangement of a typical septic system installation.

be called *leach* pipes, *drainpipes*, *absorption field* pipes or some similar term, can spread out into the absorption field in a number of different patterns.

This is essentially how the system works. Waste material, both solid and liquid, travels into the house drain, then the house sewer line and into the septic tank. The raw sewage decomposes by bacterial action within the tank, with the solids turning to liquid and quantities of gases being generated. A small amount of solid particles falls to the bottom of the tank and becomes sludge. Each time a quantity of raw sewage runs into the inlet of the tank, a similar quantity of liquid is displaced to drain out through the outlet. This *effluent* travels down the outlet sewer pipe and into the distribution box and/or the disposal pipe system. If the system is properly constructed, the effluent will discharge approximately equally into the disposal lines and drain away into the soil around them. Much of the effluent is carried away through the soil, while some disappears by evaporative action from the surface of the absorption field.

There are a number of details involved in septic system installation, as well as numerous designs, different possible configurations for absorption systems, various types of tank arrangements and so on. These will be investigated further in a later chapter.

The Third Alternative

Until recently, if a municipal system was unavailable and a septic system could not be installed because of nonabsorptive soil conditions, the property owner was stumped. This is still true in many areas. However, the situation is changing. Various systems have been in use for some years in other parts of the world that effectively dispose of both human and nonhuman wastes (often separately) in a most effective fashion, without using any fresh running water at all. These systems have recently been introduced in this country, and similar ones are now being designed and manufactured here. There are a number of different types and several different operational setups, and delving into them in detail is outside the scope of this book. But mention is made of them for a particular reason.

If you find yourself in this situation, investigation of one of the so-called *biodegradable sewage disposal systems*, such as the *Clivus Multrum*, might be well worth your while. More and more building departments and similar authorities are now recognizing the fact that these systems do indeed fill a definite need in certain circumstances, and they are effective, healthful, and perfectly workable arrangements. Consequently, they are now being allowed in many localities where there seems to be no other alternative.

In fact, these systems do work very well, though they are rather expensive. But in the event of their use, considerable changes must be made in the overall plumbing plans as opposed to a conventional system. For instance, the entire drain-waste-vent system must be changed around, and a good share of it is eliminated. All human waste is dumped directly into the unit, so the large soil and waste stack or pipes are eliminated. Gray-water disposal can be accomplished in a number of ways, but generally speaking the overall disposal system is more compact and employs relatively small pipes, and fewer of them. Disposal may be done by the unit itself or by elimination into a relatively small dry well or some similar arrangement, and traps may be reduced in number or even eliminated.

Obviously there is a need for some plumbing with these systems, but the amount is variable depending upon the specific type of system. The cold water supply system is modified slightly, since there are no toilets requiring fresh water for their operation. A plumbing system which will be used in conjunction with one of these disposal units should be designed with the particular unit in question as a starting point. But the basics of that design will be the same as for any other normal plumbing system; the principal differences will lie in the inclusion or exclusion of certain segments of the system, and a few detail differences in the particular fittings and pipe sizes used.

AUXILIARY SYSTEMS

There are few other plumbing subsystems which may or may not be included in the overall plumbing system for a given residence. These auxiliary systems may be interconnected with the main plumbing system, or they may exist as essentially separate entities.

Sprinkling System

One of the more common auxiliary systems today, especially with new houses, is a lawn sprinkling system. Once an adequate design is settled upon, the installation consists primarily of connecting a supply line to the house cold water system and running the line to a series of branches that in turn connect with sprinkler heads located at various points about the premises. These systems may also include accessory items like flow regulators, timers, gallonage indicators and the like. They can also be automated to turn on and off at preset times or after a predetermined quantity of water has passed through the system. Though the layout is a bit different, such systems are also used for sprinkling certain types of private tennis courts. And, of course, they may be adapted for watering vegetable gardens.

The circumstance can easily occur where a sprinkling system cannot be made to function properly because of low pressure or insufficient volume flowing through the building's cold water system. In such cases it may be necessary to install a booster pump and auxiliary storage tank at the supply end of the sprinkling system. If the water supply is derived from a private source, the answer may be to install a larger water pump and/or line. Or, if the source is easily accessible, like a shallow well or a pond, an auxiliary pump can be installed at the source and the sprinkling system supply line connected to it, bypassing the house water supply entirely.

Irrigation System

As opposed to an ordinary sprinkling system, irrigation systems are generally designed to supply large quantities of water to extensive areas, particularly big truck gardens, fields of grain, pastures and the like. Since the necessary water volume is large and the water need not be purified or even potable, such systems are not generally connected to a domestic cold water supply.

The water source may be an open running irrigation ditch or canal from which the irrigation water can be routed in small trenches or pumped into above-ground piping systems with auxiliary pumps. Water may be taken from a holding pond or reservoir by pump, or from deep wells drilled for the purpose. Wherever plumbing is required, considerable use is made of plastic piping.

Swimming Pools

Swimming pools have their own systems of pumps, filters, chlorinators, valves and assorted fittings. The bulk of this equipment is generally included in the swimming pool package and is installed and checked by the contractor doing the pool job unless, of course, you happen to build your own pool.

Pools also require a water supply line and some sort of discharge and drainage unit. The water supply line can be connected directly to the house cold water system, or can be taken directly from another convenient water source, provided that the water is acceptable in quality. The drainage system can be handled in a number of different ways. Some pools are emptied by means of a portable electric or gasoline-driven centrifugal pump. The water is discharged, if allowable, into the open to run down the nearest gutter or drainage ditch and away, perhaps into storm sewer system. Or the pool may be equipped with an integral drainage pump to accomplish the same purpose. In some instances the drainage flow may be directed into a municipal sewage system by direct piping. However, this is often forbidden because of the extra unnecessary load placed upon the sewage treatment plant.

Where neither of these possibilities are permissible, other arrangements have to be made. Small pools can be and often are drained by direct piping into the septic tank system. This is not an advisable method, though, because the large quantity of water pouring into the tank quickly serves to flush bacterial vital to the sewage decomposition process out into the absorption field, fouling up the correct operation of both the field and the tank. A much better solution, especially in view of the fact that the swimming pool discharge water is essentially pure, is to drain the water into a large separate dry well of sufficient capacity to handle that substantial volume of water. If only a small dry well can be built, its capacity can be increased by the addition of a separate absorption field, much after the manner of the septic system.

Auxiliary Drainage

There are a great many different auxiliary drainage systems that might be installed around a house. Sometimes several are used in conjunction with one another, and they may be interconnected or not. Sometimes such drainage systems are direct-connected to a municipal storm sewer system, but more often they serve only to route excess moisture away from buildings or other improvements.

For example, drainpipe networks are sometimes installed below slab-on-grade poured concrete foundations that are laid in areas of consistently high ground moisture content. The system neutralizes any potential damage that might occur to the foundation, because of excessive moisture or freezing. Private tennis courts, particularly of the porous types, are treated in just the same way.

Foundation drains are also very common; these drainpipes are installed around the perimeter of a foundation just above the footing and serve to carry away excess moisture that might otherwise penetrate the foundation walls. Similar drainpipe may be laid alongside nonporous tennis court installations, walks or driveways, retaining walls, patios and the like. Simple catch basins can be attached to short lengths of underground drainpipe and placed beneath downspouts attached to roof gutters, in order to direct collected rainwater away from the foundation.

In short, anywhere on the premises that there might be excess water, either below or above ground, a drainage system of one sort or another can rectify or at least help to minimize the problem. Though these drainage systems should not be connected to the house drain in any way, nonetheless they are sometimes important and perhaps even vital adjuncts to the overall residential plumbing system.

Heating and Fuel Systems

In order to complete the picture of a residential plumbing system, mention must be made of both heating and fuel-carrying plumbing systems. A plumbing network is necessary when installing either a hot-water comfort heating system or a steam heating system. Fuel-carrying systems include the relatively small amount of piping necessary to introduce utility gas into the building for heating, cooking and similar purposes. The piping or tubing systems used for conveying liquified petroleum gas (LPG) or the so-called "bottled gas" into a home also involve a very small amount of plumbing. The fuel oil delivery line is generally the smallest of all and consists merely of a length of copper tubing running from the fuel storage tank to the oil burner.

These plumbing systems are put together mechanically in much the same fashion as any other plumbing system, though many of the specific details and the rules and regulations governing the installations are quite different. In any case, plumbing of this ilk is outside the scope of this discussion. Most of the systems are entirely metallic, and plastic materials are just now coming into use in limited fuel-carrying applications in a few areas, under the aegis of the utility companies and governing bodies involved.

Chapter 2

Plastic Plumbing Components

Over the past few decades the various component parts that go into the makeup of a complete plumbing system have become more or less standardized. There are many different kinds of pipe and fittings available to fulfull practically any purpose. Plastic plumbing components are relatively new on the scene, but both pipe and fittings follow essentially the same format as for metallic plumbing systems. Those who are familiar with the fittings and accessory items used in making up a galvanized iron or copper plumbing system will immediately recognize the plastic counterparts. There is not, however, quite as great a variety of specialized fittings for plastic systems as for others. But, because of the great flexibility of plastic components, such specialized fittings are not generally required; methods of fitting a plastic system together can be easily managed to circumvent specialized items.

Because of the relative newness of plastic plumbing systems the advantages and disadvantages of the material, as well as specific information on exactly what materials are available to work with, are not widely known. About the best way for the would-be home plumber to recognize the value of plastic plumbing, and to realize the great variety of component parts that are available for all purposes, is to investigate the field in depth.

ADVANTAGES

Perhaps the greatest advantage of plastic pipe and fittings for the do-it-yourselfer is the unexcelled ease with which it can be worked. For instance, cutting plastic pipe is a breeze, and can be done with ordinary hand tools regardless of the pipe size. Joining plastic pipe and fittings is done by any of several methods, depending upon the type of pipe and fittings involved, but in every case the job is as simple as can be.

Weight

Weight is another important factor. While galvanized pipe in a nominal ¾-inch size weighs anywhere from a bit over 1 pound to as much as 1½ pounds per foot, chlorinated polyvinyl chloride (CPCV) plastic pipe weighs a bit less than 1½ pounds per 10-foot length. It is safe to say that on the average plastic water pipe and fittings weigh about a tenth as much as black or galvanized steel pipe. This means not only far greater ease in handling for the plumber, but also much less weight to be supported by the house structure when the pumbing system is installed. The difference between the weight of plastic drain, waste and vent pipe and cast iron drainpipe is not as great; nominal 4-inch polyvinyl chloride (PVC) pipe weighs about 2 pounds per foot, while the same diameter of single-hub service-weight cast iron pipe runs about 8 pounds per foot (some types may reach 12 or more pounds per foot). Even though the plastic pipe is only about one-quarter as heavy as the cast iron, rather than a tenth as with the water supply piping, the difference is even more marked when it comes to handling the material. Boosting a 10-foot, 20-pound length of PVC around is a lot easier than muscling a 5-foot, 40-pound chunk of cast iron into place.

Many more similar comparisons could be made between plastic pipe and fittings and various kinds of metallic plumbing materials. Nominal ½-inch polybutylene tubing, for instance, weighs only about 5 pounds per 100-foot coil, but the same length of the same size brass pipe or copper tubing would weigh roughly 20 times as much. The point is, plastic pipe and fittings weigh far, far less than any metallic plumbing components and are therefore much more easily handled and installed.

Cost

Another factor of prime interest to the do-it-yourselfer is the cost of plastic plumbing components. The cost of specific pipes and fittings vary widely, of course. Price is dependent upon a great many factors, and can only be effectively compared with other materials by checking these particular items at a given time and at the local level. The continuing crude oil crisis and the consequent shortage of materials from which plastic products are manufactured may well mean that the cost of plastic plumbing components will rise at a somewhat greater rate than metallic piping and fittings.

It is fairly safe to say that the homeowner can realize substantial savings in the cost of pipe and fittings by using all plastic parts. As an example, at the time of this writing, the cost of nominal ½-inch Type K copper tubing was approximately three times that of nominal ¾-inch flexible polyethyene plastic pipe. The differences in fitting costs were even greater. One must consider, too, the costs in time to the installer of the materials. Because the job can be done faster and easier with plastic as a general rule, time costs are also considerably less than with metallic systems.

Ease of Installation

Another point of particular interest to the do-it-yourselfer is the ease with which plastic pipe can be installed. Not only is it light in weight, but it also ranges in flexibility from limber to virtually limp. For instance, PE (polyethyene) water supply pipe, which comes in coils, is easily handled and unrolled into a relatively straight line. It is flexible enough to turn corners with relative ease and can be wiggled into place with minimum difficulty. Polyvinyl chloride or chlorinated polyvinyl chloride is classed as rigid pipe, but it, too, is relatively limber. While it cannot be run around corners, nonetheless the lack of stiffness and rigidity found in black steel pipe, for instance, makes for easier installation in most instances.

Polybutylene tubing is so flexible that it can actually be pulled through bored holes or fished through building cavities in the same manner as an electrical wire, with equal ease. This is unquestionably the easiest piping or tubing of all to install, especially in difficult areas or cramped quarters. An added feature of the partly or fully flexible piping or tubing is that long runs can be made with few and sometimes no fittings, just a solid, unbroken run of pipe. This situation is often of considerable consequence in saving time in the installation as well as costs for fittings.

Qualities of Material

Apart from the advantages that accrue to the do-it-yourself plumber by using plastic pipe and fittings, there are other advantages as well that are inherent in the material itself. For instance, most types of plastic pipe are tough and sturdy, and very forgiving of mechanical abuse (within reason) and rough handling. A hard blow with a hammer will flatten a piece of ½-inch copper tubing, but will not leave a dent in a piece of CPVC pipe. If PB tubing gets squashed flat, it returns to its former shape. A length of 4-inch drainpipe when dropped will bounce; a length of cast iron will smash.

Plastic pipe is also extremely resistant to corrosion, much more so than any metallic piping. It can be buried in damp acid soil or directly in cinders with no fear of damage, and it can be buried next to metallic pipe with no chance of galvanic action taking place to corrode the pipes. It is totally nonconductive of electric currents. Plastic does not rust, nor is it bothered by direct burial in the earth or by exposure to the air and elements. And since the material is so slick and smooth, practically nothing will adhere to the inside walls of the pipe. Rust does not build up; nor does lime, calcium or any other mineral that might be present in the water.

While steel pipes are prone to substantial buildups of minerals on the interior walls after a long period of time, even to the extent of choking them completely, this will never happen with plastic water supply piping. The fittings, of course, have the same characteristics. By the same token, the inside wall of drainage pipes are not prone to accumulation of residues, buildups of grease or scales and the like. This in turn means that since minor buildups are highly unlikely to occur, there is little chance for accumulations to cause pipe obstructions that could eventually lead to blockage, assuming that the pipe is properly installed in the first place.

Flow Rate

The smoothness of the interior walls of plastic pipe leads to another advantage, too. The relative absence of friction as water flows through the pipe makes for excellent flow rate characteristics, so much so that in some occasions it is possible to use a plastic water supply pipe of one diameter smaller than galvanized steel pipe for the same job. And, since there is no scaling or other buildup to hinder that flow, even over a long period of time, the characteristic excellent flow rate will remain constant for as long as the pipe is in service, something that cannot be said of most metallic pipelines.

Expandable Pipe

Flexible plastic pipe, because of its nature, is frequently an excellent choice for installation of water supply lines in areas of severe winter weather. This is because the pipe is expandable to a degree. Water freezing inside the lines merely pushes the pipe walls outward, without rupturing them. The fittings, however, will indeed split if they freeze solid. But the advantage lies in the fact that flexible plastic water supply lines can be installed just slightly below ground level or even on the ground surface, and simply opened in the event of cold weather. Even though portions of the pipeline may be partly filled with water under no pressure, the pipe will not break. Metallic piping, on the other hand, will almost always split open unless completely drained.

In fact, even rigid plastic water supply piping, such as PVC or CPVC, will tolerate a certain amount of freezeup in normal applications without splitting apart. As long as as the freeze does not occur at a fitting, the pipe will swell, then return to normal shape upon thawing, with no damage done. Fittings, however, will invariably split. Also, if left frozen for substantial lengths of time and in very cold temperatures, the rigid pipe itself will shatter. There is a good possibility that no damage will occur from a short accidental freezeup in a residential plumbing application. Practically any kind of freezeup, no matter how short, will cause damage in a metallic system. Thus, plastic piping is likely to be a better bet for installation in winter climes.

Insulation

There is another interesting characteristic of plastic piping, and that is the fact that the material itself is an insulator. This means two more plusses, though perhaps relatively minor, for the owner of a plastic plumbing system. First, hot water flowing through a plastic pipeline loses much less heat to the pipe walls and into the air than it does in a metallic pipeline. Hot water gets from the tank to the tap faster and with less heat loss, which mean somewhat lower hot water usage. Also, the exterior walls of plastic pipes seldom become hot enough to the touch to cause pain or a burn, so exposed plastic hot water pipes constitute no safety hazard for youngsters. The second plus lies in the fact that cold water also remains insulated from the outside air, and as a consequence plastic water pipes almost never sweat or drip from condensation as metallic cold water piping frequently does in damp, humid and warm climates. This means no puddles to clean up, no mildew, and no dry rot potential within the building framework.

There is one more item, too. Because of the inherent properties of the material and the flexibility of both the plastic pipe and the complete plumbing system, sound waves are quite effectively damped at or near the source, rather than being transmitted throughout the system as is often the case with metallic piping.

DISADVANTAGES

The long list of advantages that accrue not only to the do-it-yourselfer but also to the homeowner fortunate enough to have a plastic plumbing system might lead one to believe that the material is a plumbing paragon that has no disadvantages. But nothing is perfect in this world, and there are indeed a few. For the most part, though, they are relatively minor, and can largely be viewed as problems or difficulties rather than major disadvantages.

Bureaucratic Regulation

Perhaps the most serious general drawback to plastic plumbing systems has nothing to do with the material or the system itself, but is rather a matter of bureaucratic regulation. Unfortunately, there are still many areas in this country where plastic plumbing components are disallowed and cannot be installed. This is a highly variable situation that depends entirely upon local building and/or plumbing codes. In some areas no use can be made of plastic at all. In others, plastic is in common use for such purposes as septic leaching field drainpipes, ordinary ground or rainwater drainage systems and the like. In other locales it can be installed for those purposes as well as residential drain, waste and vent systems. Sometimes plastic can be used for the house water supply line, but not for water distribution within the house. Some of the various rules and regulations seem to have little point or rationality behind them.

In areas where zoning regulations are not in effect, most probably building codes are also nonexistent. You can use whatever plastic material you see fit for part or all of your plumbing system. Sometimes where building codes are in effect, plastic plumbing materials are not specifically mentioned simply because the subject has never arisen. In this case, applying for a variance and/or working with building inspector and other officials to have plastic plumbing systems written into local codes as an allowable use may be a worthwhile procedure. In any event, always check with local authorities before going ahead with your plumbing plans.

Unavailable Plumbers

In many areas a somewhat similar situation arises that effectively excludes the use of plastic materials in plumbing systems. This occurs when the homeowner wants such a system installed and is not interested in doing the work himself, but cannot find a local plumber to do the job, either. Many professional plumbers simply will not install plastic materials, for a variety of reasons that as far as they are concerned are perfectly valid. Others are happy to install plastic drain, waste and vent systems, but will not put in a plastic water supply system. This situation is slowly changing, and eventually most plumbers will doubtless be happy to install complete plastic systems. Currently, however, the homeowner's plans may be thwarted, at least in part and depending upon the local attitude toward plastic plumbing systems.

Routine Installation or Misapplication Problems

As to the plumbing system itself, there are a few minor potential drawbacks or problems that might arise, but almost always they can be traced back to improper installation, misapplication of pipe or fittings or poor workmanship. This same situation, of course, applies to any plumbing system or other mechanical systems, and has nothing to do with the material itself.

As an example, if rigid plastic pipe is run through tight-fitting bored holes in joists, the lines will inevitably squeak and chatter as they expand or contract when water runs through them. But if the pipe is properly supported by special pipe hangers made for the purpose, or otherwise suspended so that the material is free to expand and move without hinderance, the situation will not occur.

Permanent Installation

Another potential drawback lies in the fact that once the system is made up, the installation is permanent and cannot be readily disassembled. With metallic piping systems, additions, revisions or corrections of mistakes are often a matter of unscrewing or unsoldering the fittings, making the necessary rearrangements and putting the whole thing back together again. With plastic systems that use welded joints, the fittings cannot be reused. On the other hand, there are usually ways to get around that problem, so that some fittings can be reused with the addition of couplings or other fittings. On the whole, the job of correcting mistakes, making additions or revising parts of the system is much simpler than with metallic piping. More will be said about this later.

A common circumstance with metallic pipelines installed in cold-weather areas is to wrap exposed, poorly positioned or uninsulated pipe with heat tapes and pipe insulation to prevent freezeups. Normal heat tape cannot, however, be wrapped around plastic piping of any sort as a permanent installation. Over a period of time, heat from the tape will deform the plastic, and eventually the pipe or fittings will fail. Should a heat-taped pipe be emptied of water for some reason, the heat tape can deform and melt the plastic in a matter of hours, ruining that portion of the system. There is one kind of highly specialized and very expensive heat tape, difficult to obtain and not stocked by hardware stores or lumberyards, that will work on plastic pipe. Relative unavailability means that in effect plastic pipes cannot be heat-taped.

Thawing Difficulties

By the same token, plastic piping systems are more difficult to thaw out if they do freeze than are metallic systems; different methods must be used. A DC welder cannot be clamped to plastic pipe to effect a thaw, because the plastic is an insulator and no current will flow. A torch or other open-flame heat cannot be used because the plastic will deform or melt. A low-power heat tape can be wrapped around a frozen pipe to effect a thaw provided that is carefully watched so that deformation does not take place. A heat source such as a sun lamp or electric unit heater can also be positioned close, but not too close, to the frozen section of the line, and it will eventually thaw out. Wrapping rags around the pipeline and soaking the rags with hot water is another possibility.

The difficulty in thawing frozen plastic pipes and the fact that they cannot be successfully wrapped with heat tape means that the plumber should take all precautions in installing pipelines, whether water supply or drain, in locations and in such a way that they are not susceptible to freezeup, even under severe conditions. Of course, in theory the same holds true for any plumbing system, regardless of the specific material used.

Susceptibility of Plastic to Heat

The susceptibility of plastic to heat can create other difficulties if an improper installation is made. For instance, not all kinds of plastic pipes are capable of carrying hot water. The types used for this purpose are CPVC or PB, and they are designed to carry domestic hot water supplies at a maximum temperature of 180 degrees Fahrenheit. Preferably the water temperature should be less than that, perhaps on the order of 160 degrees Fahrenheit. In the interest of economy and energy conservation, however, many homeowners will find that a domestic hot water temperature of 120 degrees Fahrenheit or even a bit less is ample for household purposes; any water heater can be set to that temperature. Water temperatures of more than 180 degrees Fahrenheit, or hot water run in plastic piping not designed for that purpose, will result in eventual pipe failure. And by the same token, plastic piping should never be run in close proximity to high-heat producing equipment or appliances, in order to prevent heat damage to the material.

Air Temperature

In some instances, the ambient temperature of the air in the working area where a plastic plumbing system is being installed can be problematical. Metallic systems can be put together in any air temperature and under virtually any weather conditions that are workable. Plastic systems that are solvent-welded, however, should be made up under dry conditions in air temperatures no lower than about 50 degrees Fahrenheit. Otherwise, the weld will not cure properly. Where a system must be put together in temperatures lower than this, there is a special welding solvent that can be used in freezing temperatures with some types of pipe and fittings. However, the most satisfactory solution for both the workmen and the system is to do the job under good working conditions, at least from a weather standpoint.

Pipe and Fitting Incompatibility

Incompatibility of pipes and fittings is another problem that sometimes arises and leads to difficulties with a completed system. Again, this is not the fault of the plastic plumbing system itself, but rather the result of a lack of knowledge on the part of the installer. There are several different manufacturers of plastic pipe and fittings, and several different chemical compounds that go into the makeup of the plastic material.

Two potential difficulties arise in this situation. One is that the pipe from one manufacturer may not properly seat in the fitting from another manufacturer or vice versa. This can lead to joints that are too tight to fit properly, or too loose for a correct and full-strength welded joint. The second problem is one of plastic material in parts or fittings that are incompatible to one another and will not weld up properly or an incompatibility of the welding solvent with the plastic materials being employed. The solution to both of these problems is simple enough; always use pipe and fittings for a particular system that are manufactured by only one company, and always use the correct fittings and/or welding solvents and cleaners for the particular materials being used. Substitutions are unwise, and can lead to problems.

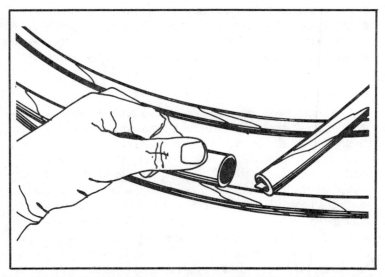

Fig. 2-1. Polyethylene (PE) pipe comes in coils and is tough and flexible. Plugged end is to keep dirt out during storage and installation.

Other so-called disadvantages or poor qualities of plastic plumbing systems are either myths or ancient history. True, during the early stages of development there were some consequential drawbacks, such as the inability of pipe of those early days to carry hot water. However, all such problems have long since been overcome, and today there is little if any argument that can be reasonably sustained against the residential application of plastic pipe and fittings for the entire plumbing system. Obviously the system must be correctly designed, the proper pipe and fittings used for the individual applications, and the system must be correctly put together in a workmanlike manner, as must any other kind of plumbing system. Provided that this is done, in the final analysis the plastic plumbing system is as good, and in the main a substantial degree better, than any other type of plumbing system.

WATER SUPPLY SYSTEM PARTS

The plastic plumbing system is made up of a considerable number of components. Several different classes of items are used, as for the water supply system, drain, waste and vent system, septic system, sprinkling or irrigation system and so on. In order to build up a proper and effective system, the right parts must be used in the right places, and all must be compatible with one another. First we will look at water supply system parts.

Polyethylene Pipe and Fittings

One of the most common plastic pipe types used for water supply is flexible polyethylene or PE (Fig. 2-1). There are several grades and sizes available, but that most commonly used for a water supply for domestic purposes when tapping into a water main, is approved, nominal ¾-inch or 1-inch diameter. The same pipe in larger sizes is widely used to supply water from wells of all kinds, and is strong enough to support the weight of a submersible pump, often without the use of a safety cable. Several pressure ratings are available, with 100-pound and 16-pound being commonly available. If polyethylene pipe is to be used to supply domestic water to a residence, make sure that the particular pipe you choose is approved by the National Sanitation Foundation for carrying drinking water. Either approved or unapproved pipe can be used for lawn sprinkling or irrigation purposes. Polyethylene is widely used in lawn sprinkling systems in a nominal ½-inch size because it is so light in weight, easy to handle and inexpensive. This pipe comes in

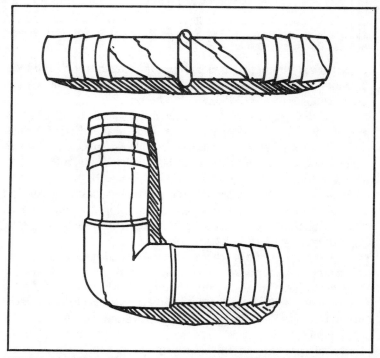

Fig. 2-2. Insertion fittings for PE pipe. Above is the coupling. Below is the 90 degree elbow.

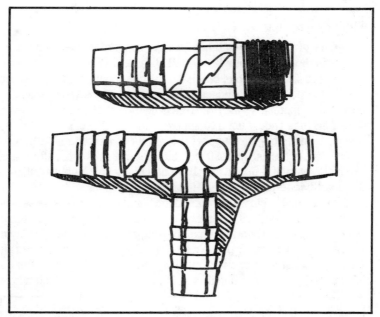

Fig. 2-3. Insertion fittings for PE pipe. Above, PE to standard pipe thread male adapter. Below is the tee.

coils that vary in length depending upon the manufacturer; 100-foot, 150-foot and 220-foot coils are a few of the standard lengths, but some suppliers will sell PE pipe by the foot.

There is only a small selection of fittings available for use with polyethylene pipe, and they may be made of metal or of molded nylon; the latter is preferable, but not always available. Straight couplings are used to join two sections of pipe, and a 90-degree elbow to make an abrupt corner (Fig. 2-2). Since the pipe is flexible, slight direction changes can be made by simply bending the pipe in a relatively wide arc. Tee fittings are available for attaching a branch line to a main line, and a male adapter makes it possible to join plastic pipe to metal pipe (Fig. 2-3).

A few other fittings are available as well, such as reducers that allow stepping up or down one pipe size. These insertion fittings, as they are called, slip inside the pipe and are held in place not only by the rings on the fittings, but also by stainless steel clamps secured around the outside of the pipe. Though this is not the only method of joining polyethylene pipe, it is the most common and least problematical approach for the do-it-yourselfer, and the fittings are readily available.

Polyvinyl Chloride Pipe and Fittings

Another commonly used pipe type for outdoor cold water supply use is polyvinyl chloride, or PVC (Fig. 2-4). This is classed as a rigid pipe, though it is in fact somewhat limber and will easily follow slight directional changes or irregular trench-bottom contours, provided they are not too abrupt. This pipe generally comes in 10-foot lengths and can be bought by the length or by the standard bundle. Sizes start at nominal ½-inch diameter, with ¾-inch and 1-inch being the most popular for residential use. For potable water supply uses, make sure that the pipe you choose is approved by the National Sanitation Foundation to safely carry drinking water.

Pressure ratings vary with this kind of pipe, and are often dependent upon the pipe diameter. In any event PVC will handle with ease any pressure likely to be placed upon it in a domestic water supply application. This type of pipe is not used with submersible well pumps, but rather to supply water from water mains, nonsubmersible pumps, reservoirs or cisterns and the like. It should only be used for cold-water applications, and never for hot water. Nor is this particular type of PVC employed for drainage purposes, and it should not be used in locations where it might be subject to freezing. It can, however, be used for irrigation purposes, but is less practical than PE pipe.

Though several methods can be used for joining PVC water supply pipe and its fittings, the most common method in residential

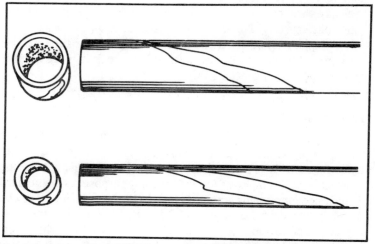

Fig. 2-4. Above, ½-inch PVC rigid water supply pipe. Below, for comparison, ½-inch CPVC pipe. Note difference in actual diameters.

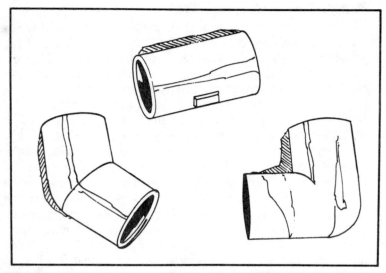

Fig. 2-5. Polyvinyl chloride pipe fittings. Left to right: 45 degree elbow, coupling, 90 degree elbow.

applications is to join the pipe with solvent-welded slip-on fittings. A modest assortment of fittings is available, sufficient to allow ample flexibility of connections and pipe runs. Lengths of pipe are joined with a coupling, and directional changes can be made with either 90-degree or 45-degree elbows (Fig. 2-5). Tees are available, as well as either male or female adapters, to join PVC pipe to metallic pipe or fittings (Fig. 2-6). Several sizes of bushings are available that allow stepping up or down one pipe size. Except for the threaded portions of the male and female adapters, these fittings are secured to the pipe with a special PVC solvent.

Polybutylene Tubing

Polybutylene, or PB, flexible pipe is relatively new to the water supply scene (Fig. 2-7). Normally called tubing rather than pipe, this material is available in several standard sizes, but the one most likely to be employed for water supply purposes is the nominal ¾-inch size, or perhaps in some instances the 1-inch size. PB can be bought in standard coils, such as 25-foot or 100-foot. It is the most flexible of pipes, and can be easily laid in narrow trenches or in cramped quarters or areas difficult to work in. It has very high corrosion resistance and fine heat resistance, and is easy to work with.

Polybutylene tubing can be joined by socket or butt fusion, but neither of these process are likely to be undertaken by the do-it-yourselfer. For water supply pipeline purposes, PB can be joined with the same type of insert fittings as were discussed under polyethylene pipe. Stainless steel clamps can be used to secure the pipe to the fittings. A few varieties of special transition fittings which allow changing from PB tubing to other lines of pipe or fittings are also made. The range is limited and variable according to the manufacturer/supplier. Ascertain what is available in your area to work with before beginning a PB piping project.

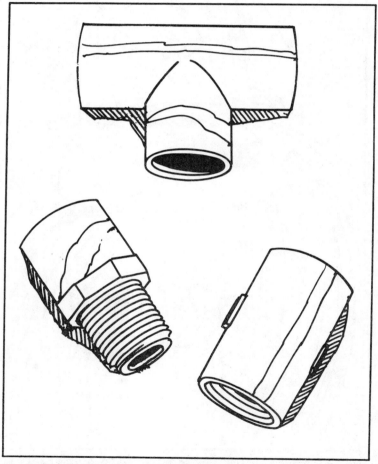

Fig. 2-6. Polyvinyl chloride pipe fittings. From left: PVC to standard pipe thread male adapter, tee, PVC to standard pipe thread female adapter.

Fig. 2-7. Polybutylene (PB) tubing being cut from coil. The PB is also available in rigid lengths.

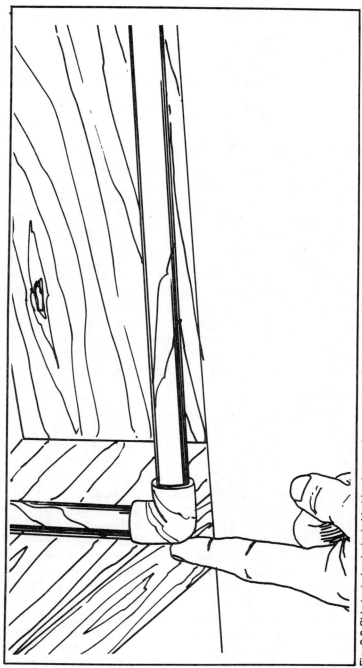

Fig. 2-8. Chlorinated polyvinyl chloride pipe can be used for either hot or cold domestic water distribution.

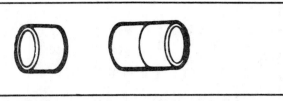

Fig. 2-9. Coupling and cap of the type used with CPVC pipe.

WATER DISTRIBUTION COMPONENTS

The type of plastic piping in most common use today for residential hot and cold water distribution systems is chlorinated polyvinyl chloride, called CPVC for short (Fig. 2-8). This is a rigid type of pipe, especially made for domestic water distribution purposes, and designed to withstand constant water temperatures of 180-degrees Fahrenheit maximum for indefinite periods. It is commonly available in nominal ½-inch and ¾-inch sizes, and comes in standard 10-foot lengths. As with other types of plastic pipe, CPVC is extremely easy to work with, and a complete internal water distribution system can be built from it. As you might expect from the name, it is a newer, tougher version of the older PVC pipe that has been used for water supply purposes for many years. It will readily withstand normal household water pressures (25-75 pounds per square inch), even at the maximum temperature rating.

CPVC Fittings

There is a good variety of fittings available to use with CPVC pipe. Straight lengths are joined with slip-on couplings, and caps are made to temporarily close the pipe for testing, to keep dirt out or while awaiting the installation of fixtures or appliances or a continuation of the line (Fig. 2-9). The caps can be permanently affixed in fabricating water hammer arresters or air chambers (Fig. 2-10). Reducing bushings allow going from one pipe size to another (Fig. 2-11). Directional changes are made with either 90-degree or 45-degree elbows (Fig. 2-12). Or a 90-degree street elbow can also be used (Fig. 2-13). While a standard elbow slips over the pipe at each fitting socket, the street elbow has one fitting socket while the other end is designed to slip inside the socket of another fitting. This end is the same size as the pipe itself.

A tee can be used to connect a branch supply line, and they are made in several configurations. Thus, the tee may be ½-inch nominal at each socket (½-inch by ½-inch by ½-inch), ¾-inch at each end and ½-inch at the tee or branch socket (¾-inch by ¾-inch by

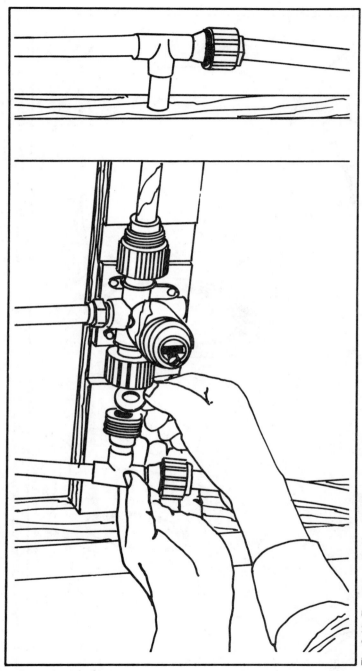

Fig. 2-10. Expansion chamber made by extending and capping CPVC supply line.

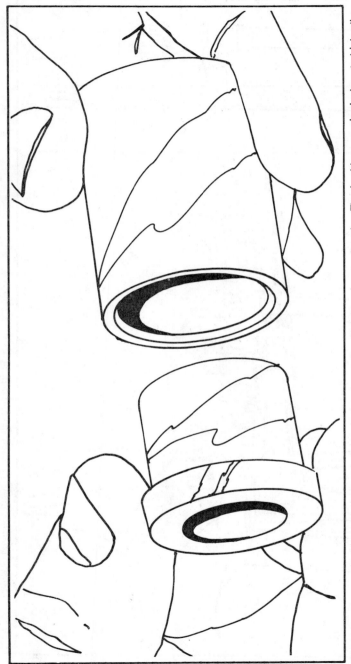

Fig. 2-11. Special reducing bushing (left) can be used with coupling (right) to change pipe size. The bushing can also be inserted into other fittings.

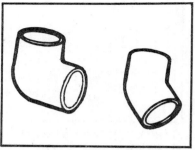

Fig. 2-12. Here are CPVC 90 and 45 degree elbows (courtesy of Genova, Inc.).

½-inch), or ¾-inch at one end and ½-inch at the other and at the tee or branch socket (¾-inch by ½-inch by ½-inch) (Fig. 2-14). In addition to these standard fittings, one can also obtain unions for direct welding to CPVC pipe, so that the piping assembly can be disassembled easily without cutting (Fig. 2-15). Stop valves of plastic can be set directly in the pipeline, and these can be obtained in either straight or angled patterns (Figs. 2-16 and 2-17). Besides providing a proper joint that compensates for differences in coefficients of expansion between the two dissimilar materials, this fitting also allows easy disassembly, as for a water heater application.

A boiler drain, similar to a sill cock or hose bib, can be directly welded to CPVC pipe to provide connections for garden hoses or automatic washing machine hoses (Fig. 2-18). A special wing elbow fitting, which has mounting ears with screw holes, can be used with the boiler drain or similar applications such as the transition from a vertical shower supply riser to the shower arm (Fig. 2-19). Escutcheons, the round plastic trim plates that cover the ragged holes in a wall through which pipes protrude, are available in two types. One simply slips over the pipe (or can be slit and placed over an existing pipe) and rests against the wall. The torque type is provided with screw holes and can be slipped over and welded to the pipe against the wall and anchored in place to prevent any pipe movement (Fig. 2-20). In addition to this interesting collection of fittings, there are

Fig. 2-13. A CPVC 90 degree street elbow (courtesy of Genova, Inc.).

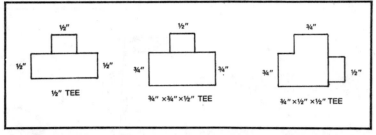

Fig. 2-14. Typical tee arrangements. The barrel size of the tee is read first, then the branch.

also various special transition adapters available to connect PB tubing to CPVC piping. These may be angled or straight, or include stop valves. Genova also produces a unique stackable tee that is used for making manifolds (Fig. 2-21).

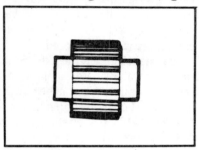

Fig. 2-15. A CPVC union fitting (courtesy of Genova, Inc.).

Small-Diameter PB Tubing

Another excellent possibility for piping domestic water distribution systems is PB tubing, the same as was discussed under water supply lines but in smaller diameters. The nominal ½-inch

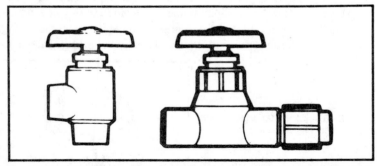

Fig. 2-16. The CPVC all -plastic angle stop valve and straight or line stop valve (courtesy of Genova, Inc).

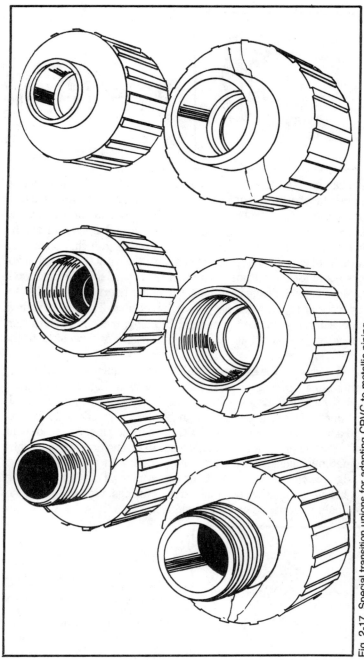

Fig. 2-17. Special transition unions for adapting CPVC to metallic piping.

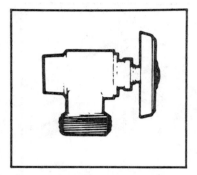

Fig. 2-18. A CPVC all-plastic boiler drain (courtesy of Genova, Inc.).

size can be used to connect fixtures or appliances that use sizable quantities of water, such as a shower or an automatic washing machine, but the ¼-inch size works perfectly well for low-volume fixtures like toilet tanks or lavatories. An entire system can be plumbed by using the three common sizes of ¾-inch, ½-inch and ¼-inch tubing, making connections with the various special adapters made for the purpose. The tubing is completely flexible, so it can be snaked through partitions, run under floors or otherwise set any place in any way that happens to be feasible. The cost of fittings is considerably reduced by using PB tubing, since full-length runs can be cut from a 100-foot coil and fittings are only required at the connection end.

Because of the extreme flexibility of PB tubing—¼-inch can be turned to a minimum radius of about 2 inches, ½-inch to 4 inches, and ¾-inch to 6 inches—fittings for directional changes obviously are unnecessary, and none are available. Connections to fixtures and appliances, or joints to CPVC or metallic pipe, are made with one or another of the appropriate special transition fittings, which sometimes must be used in combination with yet other fittings designed for use with CPVC or metal pipe. This is a matter of "cobbling up" a suitable combination of fittings to get the job done, and only experience and a knowledge of exactly what fittings are

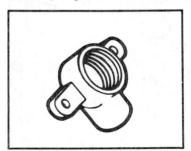

Fig. 2-19. The CPVC wing elbow attaches directly to the wall. It can be used to terminate a CPVC water line in a boiler drain, sill cock or other valve (courtesy of Genova, Inc.).

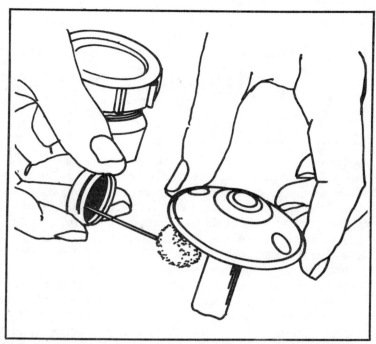

Fig. 2-20. Torque escutcheon welds to CPVC pipe stub, covering ragged hole edges in wall and anchoring the pipeline securely.

available and how they can be used will enable you to make up such combinations.

For example, if a coupling must be made between two lengths of PB tubing, the easiest method is to attach a straight adapter or transition fitting, PB at one end and CPVC at the other, to each PB length. Solvent-weld a short stub of CPVC pipe between the two adapters (Fig. 2-22). If a tee is necessary for a branch PB line, use three adapters, three stubs of CPVC and a CPVC tee fitting. By using enough adapters, pipe stubs or nipples and other fittings virtually any combination of directional changes, branches and adaptations to other systems can be made.

DRAIN-WASTE-VENT PIPING

Perhaps the most commonly used and most effective plastic piping for drain-waste-vent (DWV) systems is polyvinyl chloride or PVC (Fig. 2-23). Though the material is the same as is used for cold water supply lines, the design of the fittings is entirely different and the pipe is of a different grade and weight. Polyvinyl chloride pipe for DWV purposes is available in diameters of 1½, 2, 3, 4, and 6 inches,

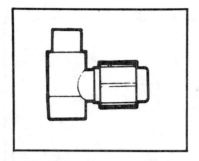

Fig. 2-21. Chlorinated polyvinyl chloride tees can be joined to one another and can accept PB tubing for making manifolds (courtesy of Genova, Inc.).

and in standard lengths of 5, 10 and 20 feet. High quality PVC-DWV pipe is the best that one can use in a system of this sort. The pipe is extremely tough, virtually impervious to corrosion from chemicals, unaffected by acids or boiling water and totally weatherproof.

The most powerful drain cleaners will not touch top quality, heavy-walled PVC-DWV; nor will the material support flame. And, like other plastic piping materials, PVC-DWV is very easy to install. There are exceptions, but normally the 1½-inch size is used for drain lines from relatively low-volume appliances and fixtures such as sinks, washbasins and the like, while the 2-inch size is best for shower drains. The 3-inch diameter is commonly used for waste pipes, waste or vent stacks and house drains. Sometimes, however, the 4-inch is used for the latter purpose.

PVC-DWV Fittings

There is a great variety of fittings available for PVC-DWV pipe, sufficient to meet practically any demands. Straight lengths of pipe are connected with couplings; reducing couplings can be employed

Fig. 2-22. One way to make up a coupling for PB lines is with a short piece of CPVC and two transition adapters (courtesy of Genova, Inc.).

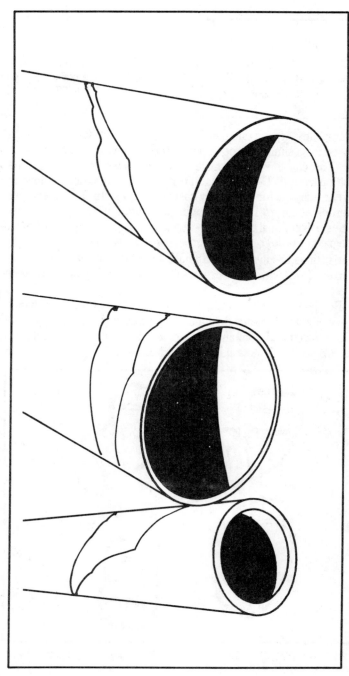

Fig. 2-23. Examples of PVC-DWV pipe. Left, standard heavy-walled 1½-inch. Center, Genova special Schedule 30 thin-walled 3-inch. Right, standard heavy-walled 3-inch. Note obvious difference in wall thickness of 3-inch sizes, as well as the outside diameters.

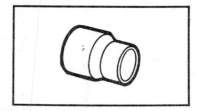

Fig. 2-24. A PVC-DWV reducing coupling to change drainpipe sizes (courtesy of Genova, Inc.).

to reduce or increase one pipe size (Fig. 2-24). Reducing bushings are used for a similar purpose but under somewhat different conditions. You can use either male or female adapters to make the transition from PVC-DWV to standard pipe threads, or a somewhat different adapter for joining to cast iron pipes. In fact, there are numerous adapters that will cover just about all transitional possibilities. Directional changes are made by using a variety of elbows: 90-degree, 60-degree, 45-degree, or 22½-degree (Fig. 2-25). These fittings are available with hubs or sockets at each end, or in the street form with a hub or socket at one end only and the other end pipe-size.

For joining branch drain lines to main lines there is a whole battery of special fittings designed for proper drainage flow, available in a variety of combinations. A *wye* is used for connecting a branch pipe into a horizontal drainpipe, or for the insertion of a

Fig. 2-25. The PVC-DWV 90, 60, 45 and 22½ degree elbows (courtesy of Genova, Inc.).

Fig. 2-26. A PVC-DWV wye fitting. Also available as reducing wye, with branch opening smaller than run openings (courtesy of Genova, Inc.).

cleanout plug in a drainpipe (Fig. 2-26). A reducing wye accomplishes the same purpose, except that the incoming branch line is smaller than the main line. In both cases the branch line enters the main line at a 45-degree angle. Where the branch line must join a vertical drain pipe at a 90-degree angle, the fitting to use is a sanitary tee when all parts are of the same size, or a reducing sanitary tee when the incoming branch line is smaller than the main line (Fig. 2-27). Both kinds of both the wye and the tee are also made in double form, so that two branch lines can join the main line from opposite directions (Fig. 2-28). A double tee with double side inlet provides four connections for branch lines, two of smaller size than the main line and two more of the same size (Fig. 2-29). Waste and vent tees are available with either double side inlets or single side inlets for further flexibility in connecting 1½-inch or 2-inch branch vent or waste piping to a vertical drainpipe, along with horizontal piping of the same size (Fig. 2-30).

There are also a number of special waste and vent fittings that are designed to allow the complete hookup of a single compact

Fig. 2-27. A PVC-DWV sanitary tee. Note down-curved branch entrance. Also available in reducing type, with branch opening smaller than run openings (courtesy of Genova, Inc.).

Fig. 2-28. A PVC-DWV double sanitary tee. Also available in several reducing sizes (courtesy of Genova, Inc.).

bathroom, and others may be obtained to connect two complete back-to-back bathrooms. Various other accessory fittings, such as closet flanges for connecting toilets, floor drains and shower drains, special capped roof drain assemblies, drum traps and P-traps, and bell traps with replacement grates are also available for use as floor drains (Fig. 2-31). In short, whatever the particular plumbing hookup situation, there is a fitting, or combination of fittings, that will do the job.

Genova Schedule 30 Pipe and Fittings

Mention must be made here of a unique DWV piping system made by Genova, fully approved by all high-level authorities (but not necessarily by local authorities) and designed particularly for residential applications and especially for the do-it-yourselfer. This system, called Schedule 30 In Wall PVC-DWV pipe and fittings, is made to fit within a conventional 2 × 4 stud wall without any structural alterations. The problem with standard DWV systems is that the large drain piping must be enclosed in a wall that is either

Fig. 2-29. A PVC-DWV double sanitary tee with double side inlets (courtesy of Genova, Inc.).

Fig. 2-30. A PVC-DWV waste and vent fitting (courtesy of Genova, Inc.).

furred out to provide clearance or built of 2 × 6s, and other structural modifications must also sometimes be made in order to get the piping through. While 3-inch nominal drainpipe is generally considered appropriate for residential application in waste drains, soil stacks and vent stacks (this is the minimum diameter), not even this size of standard DWV pipe will fit within a wall. It misses by a fraction of an inch. Genova Schedule 30 pipe, on the other hand, is

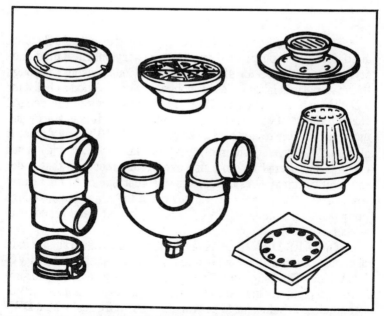

Fig. 2-31. Many specialized PVC fittings are available. Top from left: closet flange, floor drain, shower drain. Bottom: drum trap, P-trap, roof drain, bell trap (courtesy of Genova, Inc).

specially made with thin walls so that the pipe outside diameter fits almost exactly within the depth of a conventional stud wall cavity. This system is available only in nominal 3-inch size, since if a 4-inch pipe diameter is required by code it will not fit into a standard stud wall. Modifications must be made in any case.

The range of fittings available for Schedule 30 DWV pipe is less than for standard types, but the styles follow the same pattern as just discussed—couplings, adapters, elbows, sanitary tees and elbows, closet flanges, special waste and vent fittings and the like. All of these fittings are designed to accommodate Schedule 30 pipe in the main lines, but those fittings that provide for the connection of additional branch lines of smaller size are made to exactly fit the heavier-walled PVC-DWV pipe of Genova's standard-weight Schedule 40 PVC-DWV line. By using the thin-walled material wherever the 3-inch size is needed and space is at a premium, and filling out the rest of the system with standard thick-walled pipe and fittings in the smaller sizes, the do-it-yourselfer can install a complete DWV plumbing system with a minimum of trouble and expense.

ABS Pipe and Fittings

There is another type of plastic piping and fittings that is often used for DWV purposes, and that is acrylonitrile-butadiene-styrene, called ABS for short (and for obvious reasons). Though this material is in fairly widespread use, it is not as good a choice for a DWV system as is PVC. Though ABS has many of the general characteristics of other plastic piping products, it is less sturdy and more susceptible to mechanical damage than is PVC-DWV pipe. It also tends to have considerably less resistance to heat, and certain caustics and chemicals can cause exothermic reactions that under the right circumstances can lead to eventual pipe or fitting failure. And, unlike other types, it will burn. On the other hand, though perhaps not as good as PVC, ABS is indeed being used with success.

The ABS-DWV pipe is available in the several standard diameters and lengths, and is employed in the same way as PVC-DWV pipe. The range of fittings and accessories available for use with the pipe, though somewhat less extensive than for PVC, nonetheless is sufficient to put together an entire residential DWV plumbing system. There is a full complement of tees, couplings, elbows, traps, wyes, cleanouts, bushings and adapters from which to choose. As with PVC, all of these fittings are of the slip-on, solvent-welded variety. The two different types of pipe and fittings are generally easy to distinguish from one another, since ABS is jet black while PVC is buff or beige. Neither the pipe nor the fittings should be considered as interchangeable, even though they might actually fit together.

Polypropylene Pipe

A third type of plastic polypropylene or PP, has been used for many years in chemical process and waste piping, as well as other products, and is just now beginning to make its appearance in the residential plumbing field. The reason for this is simple enough. The pipe is practically indestructible under any household circumstances (and indeed, under most others as well). In household use, polypropylene very simply outperforms all other materials, regardless of what they are.

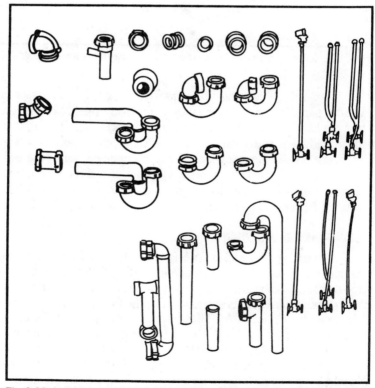

Fig. 2-32. A wide assortment of polypropylene (PP) tubular products is available for connecting fixtures to drain lines.

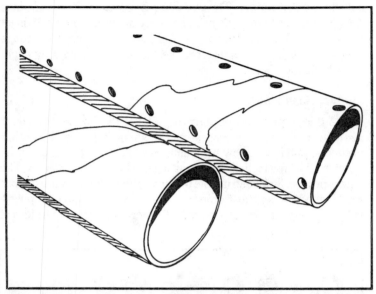

Fig. 2-33. Solid and perforated PVC pipe of the sort used for sewer and drainage purposes.

Presently PP is being offered as fixture drainpipes and traps of various sorts such as would be used under lavatories, kitchen sinks and the like (Fig. 2-32). Since traps under normal use take a lot of abuse and are prone to various sorts of failure, these polypropylene parts are a welcome innovation. They are unquestionably the best bet for installation in new plumbing systems, as well as replacement parts for old and malfunctioning parts in existing systems. In due course, PP may become available for other purposes as well.

SEWER AND DRAIN PIPING

Sewer and drain piping is designed to be used for all waste and drainage purposes in underground systems. There are four commonly used possibilities in the way of plastic piping; PE, ABS, PVC and styrene rubber, abbreviated SR. Of the four, ABS and PVC are the most readily available. Though classed as rigid pipe, there is a certain amount of flexibility in each material, especially when pipe sections are coupled into long lengths. In addition, there are some new varieties of specialized PE drainpipe that are corrugated and flexible.

Polyvinyl chloride and ABS pipe are available in standard lengths of 8 feet and 10 feet; lengths vary with manufacturers (Fig.

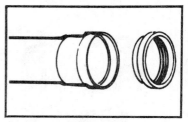

Fig. 2-34. Belled-end PVC sewer pipe with push-fit gasketed joint (courtesy of Genova, Inc.).

2-33). Though several diameters actually are available, the 4-inch size is suitable for about all residential applications and is the one most widely stocked by plumbing supply houses and lumberyards. Both types of pipe are obtainable in solid lengths for carrying sewage or drainwater, or in perforated lengths for use in leaching or drainage evaporation fields. Each type of pipe has its own set of special fittings.

Though the range of fittings is relatively small for this purpose, the nature of sewer and drainage system layouts is such that specialized or complex fittings are seldom needed. Those few regular fittings that are available are almost invariably sufficient to get the job done. Short or long elbows in either 45 degrees or 90 degrees are used to make directional changes; 22½-degree standard elbow is also available. Pipe lengths made with one belled end, or integral coupling sleeve are merely fitted together end to end (Fig. 2-34). Cut pieces, or pipe sections without belled ends, are joined with ordinary couplings. A wye or a sanitary tee is used to connect a horizontal branch line into a main sewer line. The bullnose tee is designed for making tee-joints in leaching field and seepage bed piping (Fig. 2-35). Caps can be used to close pipe ends, and crosses which accept four drain lines, are used generally with perforated pipe in seepage beds and evaporation fields (Fig. 2-36). Downspout adapters are available, as are reducing couplings, so that a 4-inch sewer line can be connected to a 3-inch main house

Fig. 2-35. Polyvinyl chloride bullnose tee used in sewer and drainage systems (courtesy of Genova, Inc.).

Fig. 2-36. Polyvinyl chloride drainage cross used in evaporation and leaching field applications (courtesy of Genova, Inc.).

drain. A special adapter is also made to join 4-inch plastic sewer pipe to clay tile lines. Another adapter makes the transition from 4-inch sewer pipe to standard pipe threads. All of these fittings are of the slip-on type, and are solvent-welded where a tight joint is desired. In some cases, however, a tight joint is not necessary and the pipe and fittings can be simply pushed together in a press fit.

Chapter 3

The Toolbox

One of the best attributes of plumbing with plastic as far as the do-it-yourselfer is concerned is the fact that so little equipment is needed to do the job. Just a handful of ordinary shop tools serves to put the system itself together. A few additional tools are needed for installing fixtures and appliances, and these too consist of common handtools. In some instances certain power tools can help speed the process, but they are the ones that most home mechanics already have on hand.

In making a full or partial plumbing system installation in new work, be that a complete new house or a new addition to an existing building, a few more tools are needed. Again, there is nothing extraordinary about either the tools or the way they are used. Remodeling, repairing or making extensive additions to an existing plumbing system generally requires the greatest array of tools and equipment. Most home shops, especially those of the more advanced do-it-yourselfers, will already be equipped with many of the necessary items.

However, there may also be a need for some specialized equipment, depending upon the amount and type of work being done. There should be no problems, though, as this equipment can't really be called uncommon, and you should be able to borrow or rent without much trouble. Buying such equipment in most cases is probably not advisable since cost can be fairly high. After the plumbing job has been completed, there will probably be little if any further use for it. If purchasing seems to be the only convenient

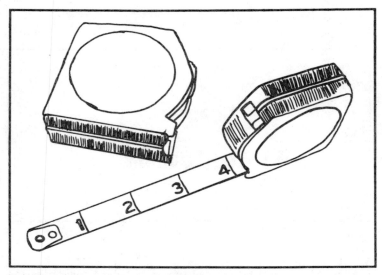

Fig. 3-1. Flexible steel tape is an essential tool.

answer, it might be possible to later sell the equipment with only a minor cash loss.

PIPEWORK

The process of actually putting together a plastic plumbing system is delightfully simple and practically effortless—so much so, in fact, that one has to remain alert against making careless or dumb mistakes. You can pick up all the tools you will need in one hand, and carry them in one back pocket or a small tool pouch.

Measuring Devices

To begin with, you will need some measuring devices. There are two of particular value. One is the *flexible steel measuring tape* (Fig. 3-1). The automatic-return belt tapes are the handiest, and these are available in standard sizes of 6, 8, 10, 12, 16, 20 and 25 feet. Pick whatever size seems to be most appropriate for the work you are doing. Actually a pair of tapes, one 8-foot or 10-foot and one 16-foot or 20-foot, seems to be about the most workable situation. For large layouts and for measuring off long lengths of pipe, like in an extensive sprinkling system, a *winding-reel tape* in either a 50-foot or 100-foot size will work nicely (Fig. 3-2).

The second kind of rule that comes in handy for this type of work is the *carpenter's folding rule* (Fig. 3-3). This is a sectional

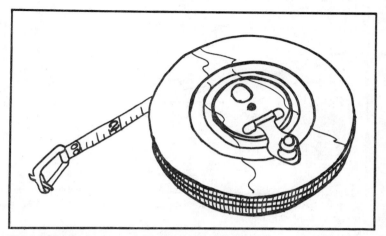

Fig. 3-2. Flexible steel or cloth winding-reel tape in 50-foot or 100-foot length is helpful for pipeline and other long measurements.

device made of wood that unfolds into a 6-foot length. The better varieties also have a small brass slide-out rule in one section, for making side measurements. While the steel tape sometimes tends to be too flexible and won't stay put or extend as far as you want it to without drooping down, the carpenter's rule is much stiffer. It is self-supporting in the vertical position and nearly so in a horizontal position, making measurements easier to take. They're also easy to use, and much more accurate than a yardstick.

Cutting Tools

The next consideration is cutting the pipe. Since plastic materials are easy to cut, just about any kind of saw that you can name will do the job. You can use a *carpenter's handsaw* of about any size

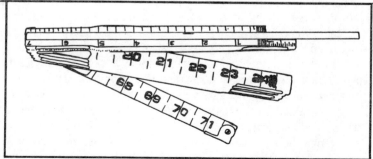

Fig. 3-3. The carpenter's folding rule with extension section.

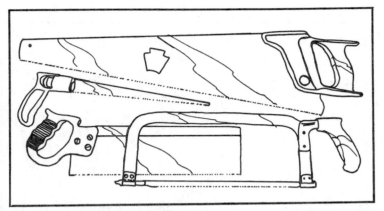

Fig. 3-4. Carpenter's hand saw, keyhole saw, back saw and hacksaw will answer most sawing needs.

or type, a *keyhole saw*, a *hacksaw*, a *backsaw*, a *miter saw* or what have you (Fig. 3-4). In all cases, though, a fine-toothed blade does a much better job than a coarse-toothed one. The cut goes more smoothly and the saw blade is less likely to catch, and the resulting cut is much less ragged and freer from burrs and chips.

Another point to remember is that some types of plastic pipe are rather abrasive, even though they appear to be relatively soft and easily cut. This has to do with the composition of the material itself. So, if you are making numerous cuts it's best not to use good blades that are primarily designed for woodworking. Instead, use an old saw or one equipped with a metal-cutting blade. A high-quality fine-toothed miter saw, for instance, will never be the same again after repeated use with some kinds of plastic pipe. And saws are expensive to have sharpened and even more so to replace.

Some kinds of plastic pipe, notably PE and PB, can readily be cut with a sharp knife. An ordinary *utility knife* works fine with the smaller sizes (Fig. 3-5). A sharp jackknife or sheath knife will do the job on either tubing or the small to medium diameters of pipe. Actually, you can use about any kind of knife with a sharp, hard blade that is not overly thick and wedge-shaped.

Power equipment will also make short work of cutting plastic pipe. A *motorized miter box* does a slick job on pipe sizes up to 2-inch nominal. A *saber* or *scroll* saw will easily cut the smaller sizes of pipe (Fig. 3-6). A *reciprocating* saw can be used to cut any size of pipe up to the blade capacity, provided the pipe is solidly anchored.

Since the pipe should always be cut as squarely as possible, some sort of cutting guide is usually in order. Most people find that

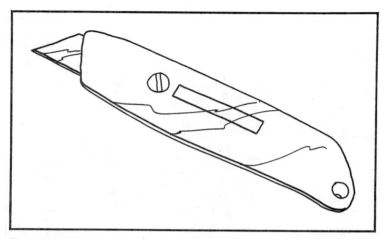

Fig. 3-5. Utility knife with sharp, flexible blade is very handy in many plumbing installations.

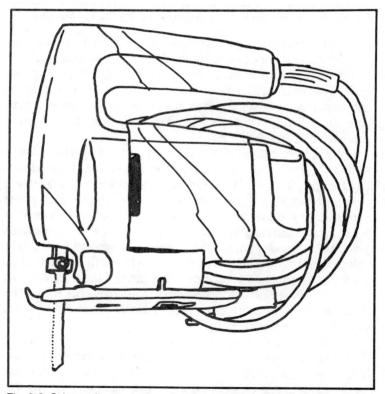

Fig. 3-6. Saber or jig saw makes short work of many cutting chores.

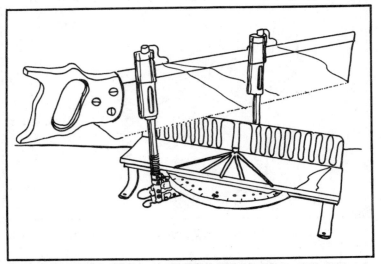

Fig. 3-7. Top-grade miter box is excellent for making squared pipe cuts.

freehand cutting of a cylindrical shape is a bit difficult, since it is so easy for the saw balde to wander and there is no guideline to follow. A *miter box* is the simplest answer for making square cuts. An expensive woodworking miter box with fitted miter saw running in guides will do the best job, but as mentioned earlier, some kinds of pipe can be tough on the saw blade (Fig. 3-7). Also, larger sizes of pipe won't fit into such a rig.

A good alternative for pipe sizes up to around 2-inch nominal is to by a cheap wood or polystyrene miter box that can be used with any handsaw. This will do a perfectly adequate job and last for a reasonable length of time at low cost, provided you do your cutting carefully. Or you can make your own miter box simply enough from wood scraps, and tailor the size of the box to fit whatever pipe sizes you will be working with. All you need to do is nail two upright sidepieces to a bottom piece. Scrap wood is perfectly adequate (Fig. 3-8). Carefully make a vertical cut down through the sidepieces, so that both cuts are exactly 90 degrees to the longitudinal axis of the box. You can do this by hand, but a table or radial arm saw will provide a more accurate cut.

The smaller sizes of pipe, like the nominal ½-inch and ¾-inch PVC and CPVC used in water supply systems, are most easily cut with an ordinary *tubing cutter* (Fig. 3-9). Most of these cutters are designed for use on copper tubing, but they work equally well with plastic. Their one problem is that the cutting wheel has a wedge-

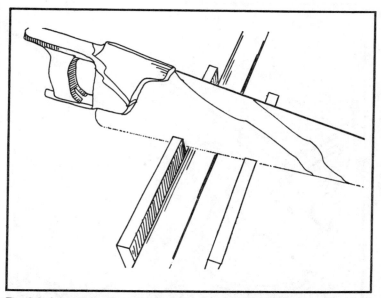

Fig. 3-8. Inexpensive but effective miter box can be homemade from three pieces of scrap wood and fitted with hand saw.

shaped blade (in cross section) that tends to leave a humped-up ridge around the edge of the cut on the pipe surface. Getting rid of this ridge is not much of a chore, but does represent one extra step. To avoid that step, you can use a cutter designed especially for plastic pipe, which has a very thin blade that does not push up a ridge. These cutters, however, are harder to find than standard tubing cutters. Whichever type you choose, you will find that this particular tool is by far the most convenient way to cut small-size rigid pipe or tubing.

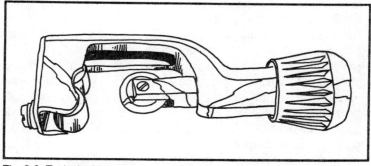

Fig. 3-9. Typical tubing cutter.

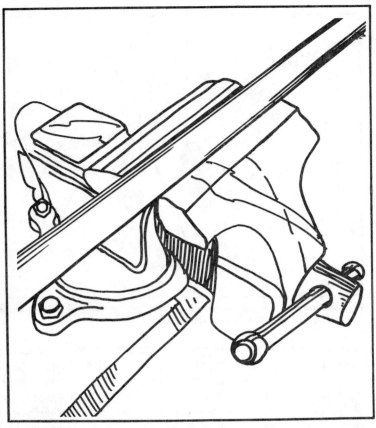

Fig. 3-10. Bench-mounted mechanic's vise can be helpful in pipe work.

Vises

When you are cutting pipe with a saw, the pipe does have a tendency to move around. Probably the most common procedure for anchoring a piece of pipe is to lay it on a sawhorse, a step or whatever else is handy and plant one's foot solidly upon it. If you have a *vise* mounted on your workbench, you'll find that a good deal easier, provided the location is reasonably convenient to the work site and you have room enough to work on (Fig. 3-10). Many bench vises are fitted with pipe jaws, and pipe can also be held between flat jaws. It is a good idea, though, to slip plastic pads over the metal jaws of a *mechanic's vise*, or to remove the metal jaw pads and replace them with wood or plastic ones. That way the pipe won't get chewed up on its outer surface. Remember, too, that you have to exercise some caution when clamping plastic pipe in a vise, as this material will crush or split open more easily (this is especially true of the thin-wall type) than will heavy steel pipe.

If you don't have a vise or you don't like the idea of using a mechanic's vise or woodworking vise for holding pipe, you might consider investing in a *pipe* vise. These are quite handy, especially if plans call for a considerable amount of cutting, and are not terribly expensive. There are two principal types. One, generally for smaller pipe sizes, uses a yoke and screw-spindle arrangement to clamp the pipe. The other is equipped with an adjustable chain mechanism. Both are available in several sizes, and can easily be attached right to the end of a sawhorse and set up anywhere that happens to be handy. They can also be used for a great many chores other than cutting pipe for a plumbing job.

Dressing Tools

No matter how carefully you cut the pipe, you'll probably find that the cut end needs a bit of dressing up before the piece can be

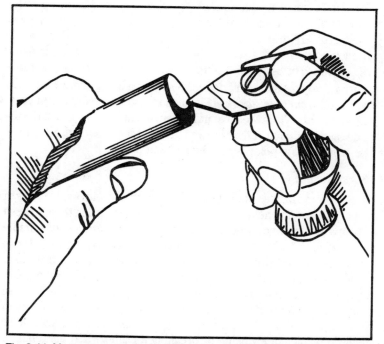

Fig. 3-11. Most tubing cutters are fitted with a reamer blade for dressing inside cut edges of pipe.

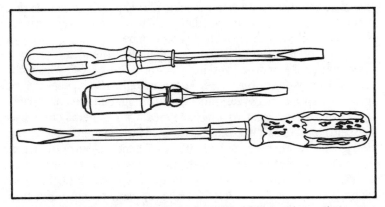

Fig. 3-12. A large screwdriver is best for tightening PE pipe clamps that secure fittings. Center screwdriver is standard 8-inch length for comparison.

put to use. There may be burrs or shavings, a ragged edge on the inside or outside, or both. Perhaps the tubing cutter has left a ridge. Whatever the difficulty, it has to be eliminated before a properly fitted joint can be made. Smaller pipe sizes can be cleaned up with a *reamer*, a triangular-shaped blade this is an integral part of most tubing cutters (Fig. 3-11). A knife scraped around the edge may do the job equally well, particularly on larger sizes. Or you can use a *file*. A 6-inch or 8-inch *single-cut* file with *smooth* or *second cut* teeth works nicely for paring off ridges. A somewhat larger *double-cut* file with *second cut* or *bastard teeth* is fine for larger sizes. In either case, a *half-round* style is the most utilitarian, because you can use the flat side on the outside of the pipe and the rounded side on the inside. A tip tapered to a point is also helpful at times.

Screwdrivers

In some phases of the plumbing system and with some types of pipe, fittings are held in place with clamps. In order to run these clamps up tight, you will need one or more *screwdrivers* of suitable size (Fig. 3-12). Even small clamps have relatively large screwheads and slots, making it easy to apply plenty of pressure. The best course is to use the largest screwdriver blade that will fit a given clamp screw. The screwdriver should also preferably have a large handle that affords a good grip, so that you can get plenty of leverage and draw the clamp up as tight as possible. Normally there is little danger of stripping or breaking a clamp when tightening it down with a standard hand screwdriver, especially if the clamp is of the high-quality stainless steel variety.

Marking Tools

As you put your plumbing system together, you'll be measuring out a great many sections of pipe to fit in here and there. Each time you do so you'll have to mark the pipe to indicate where a cut should be made. With many kinds of pipe that's simple enough—all you need is a pencil. But some types of pipe are hard and/or glossy and don't take a pencil mark well, while others are jet black and the mark is invisible. In this situation a *scriber* of some sort is necessary. An *ice pick* or a *scratch awl* works well, as does a machinist's tungsten carbide-tip steel scriber (Fig. 3-13). Even better is a chunk of fine-toothed hacksaw blade salvaged from a broken one. A short and shallow freehand cut at the desired point is perfectly obvious and easy to find on any kind of pipe.

Pouch

A container to put these tools into is handy and saves a lot of time that can so easily be lost in searching for a tool that you parked on top of a beam or left behind the step ladder. A *tool holder* or *pouch* isn't essential, of course, because you can leave the tools scattered about or you can shove them in your back pocket. However, there is a wide assortment of relatively inexpensive leather tool pouches available at nearly any hardware store, and you will find that the convenience of having one far outweighs its cost. All you need do is pick one of appropriate size and configuration for your own purposes. Once you get into the habit of slipping each tool back into its proper slot when you are finished with it temporarily, the process will quickly become automatic. The tool you want is always right at your fingertips.

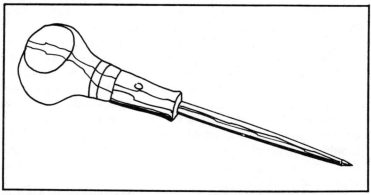

Fig. 3-13. Scratch awl is useful for both marking and hole-starting.

FIXTURE INSTALLATION

When the time comes to actually connect various fixtures and appliances, and perhaps certain valves as well, into the water system, you'll have need of a few additional tools. The same applies to a less degree to the DWV system.

There are some fixture units being sold today that have plastic attachment components of one sort or another, and perhaps are made entirely of plastic themselves. Also, various sorts of valves and stops have been perfected for direct insertion into plastic pipelines in the same manner as fittings. However, in most instances there is some sort of plastic-to-metal transition that must be made. There are a number of possibilities and many adapter and transition fittings are manufactured to meet various requirements. For the most part these fittings are solvent-welded or clamped to the plastic pipe or tubing, while the opposite ends connect to the fixtures or appliances by threaded couplings. Some can be locked down by hand, but others require the use of wrenches.

Wrenches

About the handiest kind of wrenches to use for nuts and fittings up to about 1 inch in size are *mechanic's open-end* wrenches (Fig. 3-14). These are made in specific sizes ranging from ¼-inch to as much as 1¾-inch, in 1/16-inch incremental steps. Most of them are made in double sizes; ½-inch at one end and 9/16-inch at the other, 15/16-inch at one end and 1-inch at the other and so on.

Mechanic's box-end wrenches are similar but have closed ends (Fig. 3-15). They are handy, too, but can't be used unless the nut is in the open so that the wrench can be slipped down over it. On the other hand, they do afford an excellent grip, particularly on multisided nuts. The same is true of *mechanic's socket* wrench sets, which are sometimes mighty handy to have around (Fig. 3-16).

Crescent wrenches are perhaps the most utilitarian of all, and can be used on practically anything that has at least two parallel flat surfaces (Fig. 3-17). These are also called *adjustable-end* wrenches, and as the name implies the jaws are fully adjustable through a certain range. These wrenches come in different sizes, according to both handle length and jaw capacity. The longer the handle, the greater the jaw capacity. While a 4-inch wrench might have a jaw capacity of ½-inch, a 12-inch wrench might have a jaw capacity of 1½ inches.

Another type of adjustable wrench that sometimes comes in handy is the venerable *monkey* wrench. This oldtimer does just as

Fig. 3-14. Typical mechanic's open-end and combination wrenches.

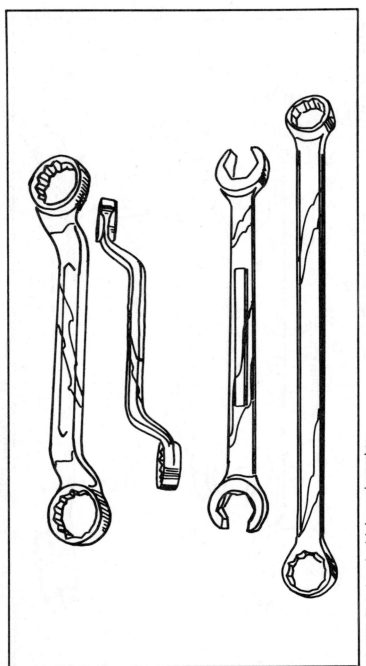

Fig. 3-15. Typical mechanic's box-end wrenches.

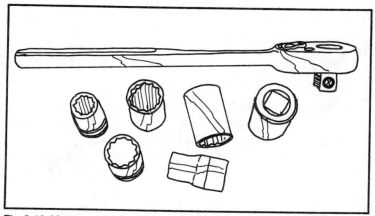

Fig. 3-16. Mechanic's ½-inch drive ratchet wrench and assorted sockets to fit.

good a job today as it did back when they were supplied in Model T Ford tool kits. Sometimes a monkey wrench can be used where others don't work as well. This wrench also is for use on flat-sided objects, and can be obtained in several sizes.

The *locking plier* wrench, more popularly known as *Vise-grips* after that well-known trade name, is also useful in some plumbing applications (Fig. 3-18). This tool, once adjusted properly, locks solidly in place and its sharp teeth take a firm hold. It has the advantage of solidly gripping objects of nearly any shape. However, it cannot be used on any object where the finish is important, because the teeth leave deep gouges and scrapes.

Adjustable-joint pliers, which are also sometimes known as *water pump pliers* or *gas pliers*, can also be useful (Fig. 3-19). These

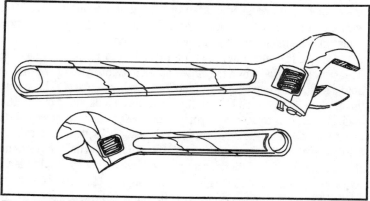

Fig. 3-17. Adjustable or crescent wrenches are valuable in plumbing work.

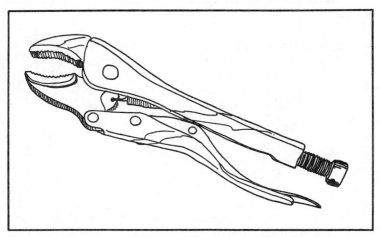

Fig. 3-18. Typical locking pliers, often known by the familiar tradename of Vise-grips.

come in numerous sizes and jaw shapes, and may be of the *arc-joint* or the *slip-joint* variety. Either type works equally well, with the arc-joint type generally having the greater capacity and affording the greater pressure and leverage capability. Like Vise-grips, they too have toothed jaws and cannot be used on finished objects.

Ordinary *utility pliers*, which may be either *slip-joint* or *solid-joint*, are also handy to have around (Fig. 3-20). There are a great many different styles of pliers, some with toothed jaws and some with flat jaws, some blunt, some rounded, some off-set, some needle-nosed and so on. You can find a pair of pliers to fit just about any requirement. Utility or gas pliers with a single-notch slip joint are probably the most utilitarian of all. They are fine for working with small fittings, final loosening of a part that already has been "started," hanging onto things and similar odd tasks.

One of the more aggravating tasks in the fixture installation department occurs when the time comes to tighten up the nuts that hold a sink, lavatory faucet or faucet assembly in place. The whole business is hidden high behind the bowl of the sink and close to the wall. It's difficult to see what you're doing. You can't get your hands or ordinary wrenches up in there. You bang your head and everything seems to be right in the way. To at least partially overcome these trials and tribulations, arm yourself with a special tool called a *basin* wrench (Fig. 3-21). This is really the only kind of tool that will do the job well, and once you get the hang of how to use one the work will progress with relative ease. The basin wrench has a

114

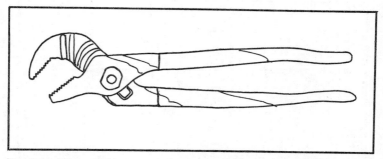

Fig. 3-19. Pliers of this type, variously known as water-pump, slip-joint or arc-joint pliers, are quite useful.

right-angle head and a spring-loaded, curved, self-adjusting jaw, all attached to a long T-handle. It is also quite inexpensive, and even though you use a basin wrench infrequently, it's a good idea to have one in the tool box.

Another area that sometimes cause difficulties in fixture installations lies with the large slip-nuts, coupling nuts or locknuts that are frequently found on various kinds of fixture drains, particularly those of tubs, sinks and basins. Though other methods can be used, the best way to tackle these nuts is with a *spud* wrench (Fig. 3-22). There are various brands, sizes and configurations of spud wrenches, and nearly all of them are adjustable. One medium-size wrench will handle the whole range of nuts that you are likely to encounter. They are not inexpensive, but they do the best job, particularly with plated or soft-metal nuts that can be easily damaged. Because the jaws are flat, the nut surface will remain unmarred. And because pressure is applied somewhat differently by the

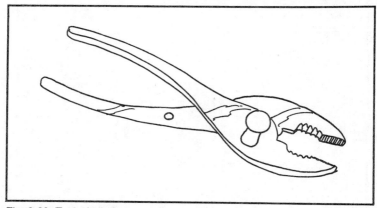

Fig. 3-20. Typical slip-joint utility pliers.

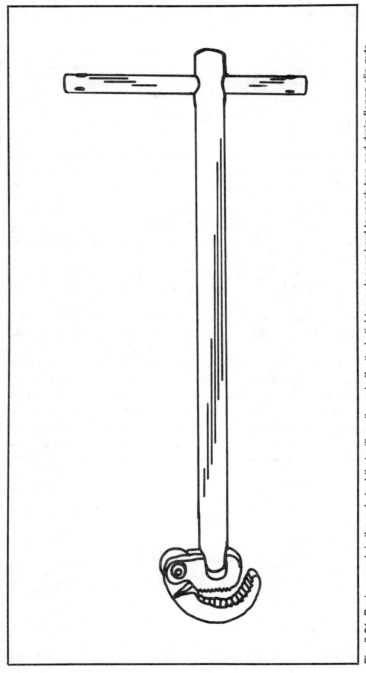

Fig. 3-21. Basin wrench is the only tool that will easily and effectively tighten or loosen hard-to-reach trap and drain flange slip nuts.

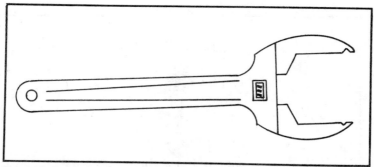

Fig. 3-22. Spud wrench, especially for use with spud nuts.

jaws of a spud wrench, the nut is unlikely to become deformed or crushed as might happen with some other kinds of large, heavy wrenches.

The *strap* wrench is another useful device, not only for plumbing but for other tasks as well (Fig. 3-23). This wrench consists of a fairly long steel handle to which a slip-strap of heavy fabric is attached. With this wrench you can work satisfactorily in cramped and crowded quarters where other types of wrenches often will not work. Also, you can wrap the strap around plated pipe and fittings without fear of harming the finish. This wrench will also grip thin-wall brass or other light metal tubing firmly with little danger of crushing.

More Screwdrivers

As you work at installing the fixtures in your system, you'll probably find that you will have to expand your selection of screw-

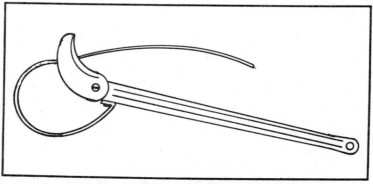

Fig. 3-23. Strap wrench protects the finish and surface of the object being worked upon.

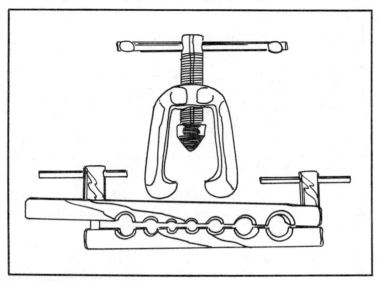

Fig. 3-24. Typical flaring tool for metal or plastic tubing.

drivers considerably beyond the one or two large ones mentioned earlier. At one time or another you will likely have need of most sizes of the standard *slotted* screwdrivers, and at least one and perhaps two sizes of *Phillips* screwdrivers. You might even find need for an offset slotted screwdriver. Ordinary mechanic's screwdrivers are fine for this purpose. Cabinetmaker's screwdrivers will also work, of course, but there is always the danger of banging them up and leaving them unfit for cabinetwork.

Flaring Tool

Though most plastic pipe and tubing is attached to the metallic parts of fixtures and appliances by means of transition or adapter fittings, there are occasions when this is not necessary. For instance, some PB and CPVC pipe and tubing can be obtained in sizes identical to that of copper tubing. In particular, the useful sizes are ⅜-inch and ½-inch nominal. Under the right job conditions the pipe can be attached to the fixture by means of a standard brass flare fitting of the type used with copper tubing. These fittings are easily obtainable at any supply house, are inexpensive, and do away with the need for a special transition or adapter fitting.

The pipe or tubing is flared in almost exactly the same manner as copper tubing, and the fitting is installed in the same way. The device used to make the flare is called, logically enough, a *flaring*

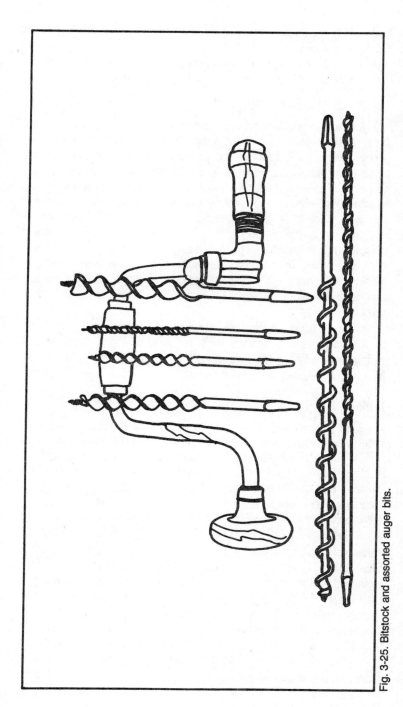

Fig. 3-25. Bitstock and assorted auger bits.

tool, and is the same one used for plumbing or automotive copper tubing work (Fig. 3-24). They are easy to obtain and not very expensive, so if the possibility exists for you to make your fixture connections in this way, you might consider investing in one.

SYSTEM INSTALLATIONS

Obviously a plumbing system must not only be put together, but it must be installed in the building. In practice, of course, the two processes go hand in hand and are inseparable; they are broken apart here simply for convenience in discussing them. The initial stages of a plumbing system installation are collectively known as *roughing-in*, which consists of putting all the pipes and fittings and valves in place while the building is still an open, unfinished structure and everything is accessible. When the roughing-in has been completed, the building is then finished off and trimmed out. Then the final phase of the plumbing installation, the *finish-work*, can take place. This consists of installing and connecting appliances, basins and sinks, showers and tubs, faucets and sill cocks, trim plates and the like.

The roughing-in process, then, will require yet a few more tools in addition to those already mentioned. Since much of the piping ordinarily travels through the structural components of the building, boring equipment of one sort or another is essential.

Drills

The time-honored strongarm method of boring holes is accomplished with a *bitstock* and an assortment of *auger bits* (Fig. 3-25). The ratcheting bitstock, or bit brace as it is sometimes called, is the most useful type. By locking the ratcheting device the bitstock can be turned in full revolutions. By adjusting the ratchet lock to one direction or the other, the bitstock can be turned in small arcs, handy in close quarters where you can't get a full swing.

The auger bits that are used with a bitstock are available in sizes from ¼-inch diameter to 1-inch, in increments of 1/16-inch. Sizes from 1-inch to 1½-inch are available in ⅛-inch increments. Standard bits are relatively short and will not bore to a great depth (the length of the bit is dependent to a small degree upon the diameter of the bit). However, there are several types of extra-long bits that are also available, like *ship augers* or *telephone bits*.

Large holes can also be drilled with a bitstock equipped with a special *expansive bit* (Fig. 3-26). This features an adjustable wing-blade on an auger shank, that can be set to any desired hole

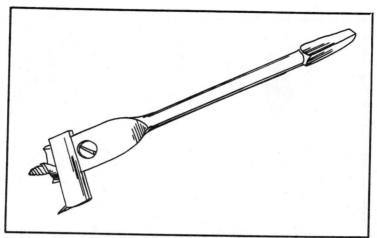

Fig. 3-26. Adjustable expansive bit for use in bitstock.

diameter within the adjustment range of the blade. There are two common ranges, ½ to 1½ inches and ⅞ to 3 inches.

If you get the impression that boring holes with a bitstock and augers might be a whole lot of work, you're exactly right. It is. But years back that was the only way, and it is just as effective today as it was then.

There is an easier way, and that's an *electric drill* (Fig. 3-27). There are three popular sizes: ¼-inch, ⅜-inch and ½-inch. The size designation refers to the bit-diameter capacity of the chuck on the drill. This is also a general indication of power, since the larger the capacity, the more powerful the drill. There are many models available ranging from cheap to excellent, and also larger sizes, although the latter are generally used in industrial and heavy commercial applications.

The ¼-inch size is easy to use because it is compact and lightweight, but its capabilities are really insufficient to make it fully useful in a plumbing installation. It is capable of boring holes up to an inch or a bit more in diameter through wood, but a great deal of strain is imposed and the tool is overworked. If you only have half a dozen holes to drill, you won't have any problems provided that you proceed slowly and keep the drill from overheating. However, hard drilling through thick material with oversized bits on a continuous basis will burn out even a high-quality ¼-inch drill in fairly short order.

Probably the best all-around, general-purpose electric drill for plumbing work (and other light construction and shopwork as well)

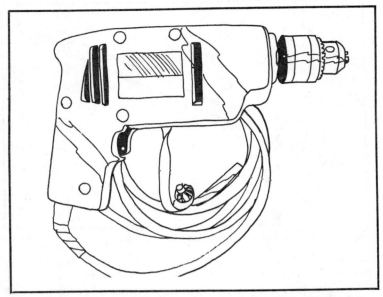

Fig. 3-27. Typical ⅜-inch electric drill, most helpful in many plumbing projects.

is the ⅜-inch size. That extra ⅛-inch of chuck capacity makes a tremendous difference. This drill is physically larger, considerably more powerful and is capable of doing relatively hard work on a constantly repetitive basis. Drilling holes of 1-inch diameter or so through wood is an easy chore, and much larger holes can be bored if the job is done carefully and the drill is not driven beyond its power limits.

For sheer power and the ability to handle just about any job the homeowner can find for it, the ½-inch drill is the choice. This rig will handle all the big jobs and is capable of tough continuous drilling in all kinds of material, including masonry. Though bulky, heavy and somewhat awkward to use, it gets the job done.

You may already have one of these drills on hand, or if not, you can doubtless borrow or rent one. But if you plan to purchase one to help you out in your plumbing installation, there are a few points to keep in mind. First, irrespective of the drill size, the best bet is to opt for a top-quality, brand-name tool of the heavy-duty variety. The very inexpensive electric drills are cheap in both senses of the word and usually cost more in the long run than the expensive models.

Second, when considering a given selection of electric drills, choose the one with the highest horsepower rating. This will give you the most powerful drill and the one least likely to burn out under hard usage, or to overload with too-great ease. If the horsepower rating is not given, look for the highest amperage or current draw rating; the higher the amperage the higher the horsepower.

Third, look for units that are equipped with ball and roller bearings that are permanently lubricated. Stay away from sleeve bearings and bushings wherever you can, because generally speaking they don't hold up as well. Also, the ⅜-inch and ½-inch drills should be fitted with a top or side assist handle to allow the user to get a firm grip on the tool during heavy drilling. A reversing feature is handy but not essential, and the variable-speed capability is also nice.

The cordless drills, incidentally, aren't too practical for this kind of work, unless you have only an occasional hole to drill. They are not overly blessed with power, nor will they run for very long before they need to be recharged.

Bits For Electric Drills

There are several kinds of bits that can be used with electric drills, and the type to choose depends largely upon the job at hand. Standard *twist drill* bits are readily available in dozens of sizes, including fractional sizes, number sizes and letter sizes (Fig. 3-28). These can be used for drilling in most nonabrasive materials. Frac-

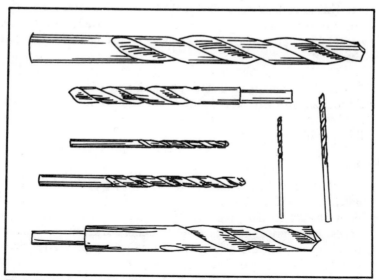

Fig 3-28. Assorted twist drill bits for use in electric drill.

tional sizes are the most commonly employed in this kind of work, and are made from 1/16-inch diameter up to 1-inch diameter in various shank sizes. But these bits are stubby and won't bore very deep.

For boring in wood in general construction and building work, the most commonly used bits are variously known as *power wood-boring* bits, *speed* bits or *spade* bits (Fig. 3-29). These consist of a flat blade with a centering and starting point integral with a ¼-inch shank, all of hardened alloy steel. They can be used in any size drill, are available from ⅜-inch diameter up to 1½-inch diameter, and cut very effectively. They are also inexpensive, and can easily be resharpened a number of times before replacement is necessary. It's possible to bore to about a 4½-inch depth with a speed bit.

When the depth of the hole must be greater than that, or when the drill must be held at some distance from the hole because of cramped quarters, there are a couple of further possibilities. One is to add a *bit extension* to a speed bit. The bit is set in the end of the extension shank and secured with a setscrew, while the other end of the shank is chucked in the drill in the usual fashion. This arrangement extends the useful length of the bit to somewhat over 16 inches. It's also most helpful in reaching into awkward spots, even though the hole to be drilled is not very deep.

Another possibility is to use a special bit called an *electrician's* bit. This is nothing more than a stretched out carbon steel twist drill, measuring about 18 inches long. They are available in several diameters, but in a much smaller range of sizes than other types of bits.

Drilling in masonry requires another type, a *masonry* bit (Fig. 3-30). These too are available in a small range of sizes and in all three shank diameters. They are specially fluted for chip removal, and are equipped with tungsten-carbide tips designed especially for this purpose.

Saws

Sometimes the occasion arises where a pipeline must be set in notches in the framing, or portions of the framework must be cut away slightly to allow passage of a drain line, for instance. The handsaws of various sorts that were mentioned earlier will take care of this chore, but the blades in this instance should be relatively coarse and of the wood-cutting variety, rather than fine-toothed metal cutting blades. A carpenter's or panel saw should have teeth in the crosscut configuration, while hacksaws, keyhole saws and the

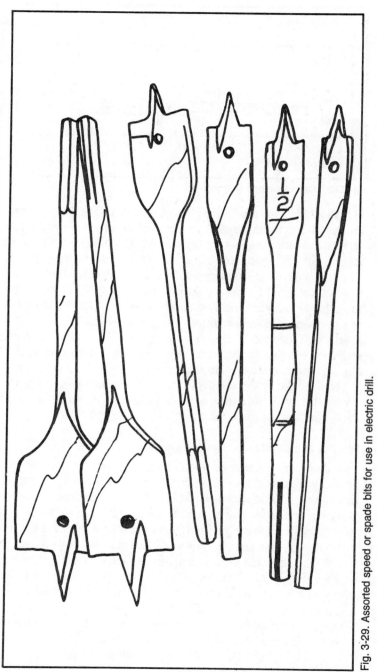

Fig. 3-29. Assorted speed or spade bits for use in electric drill.

125

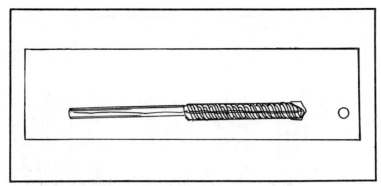

Fig. 3-30. Carbide-tipped masonry drill bit.

like should be fitted with fairly coarse blades. The keyhole saw, incidentally, is most versatile and can make cuts in cramped quarters where other saws cannot be used. It's also very handy for cutting out large-diameter holes for 3-inch and 4-inch pipes.

If you have a great many notches of similar size to cut and you are not bothered by close quarters, a power saw can do the job much faster and more easily than a handsaw. Either the saber saw or the reciprocating saw, fitted with a wood-cutting blade, works pretty well for some tasks. A *power circular* saw, also frequently called by the popular brand-name of Skilsaw, makes fast work of cutting repetitive notches in studs or rafters (Fig. 3-31). By setting the blade to the proper depth of cut, you can make two quick passes across the top of the stud, and a quick tap with a hammer splits the small chunk of waste wood away, leaving a perfect notch.

Plumbs and Levels

Running pipelines of any kind invariably involves keeping them properly aligned in one direction or another. Often it is desirable that a pipe be perfectly vertical, or plumb. Sometimes there are sufficient reference points at hand so that you can measure from various locations to the pipe and thereby make a vertical installation, or nearly so. This system, however, doesn't always work out too easily. The best way to make a determination is to use a *plumb line and bob* (Fig. 3-32). This consists of nothing more than a length of heavy string, often on a reel, with a pointed steel or brass weight attached to one end. By dangling the bob in position you can achieve a perfectly vertical line, and you can also transfer measurements or a drilling location from one point to another one directly below, and be assured that the two points will line up perfectly.

Horizontal pipelines are generally installed with a slight pitch. There are a number of ways to determine pipe pitch, and frequently they are used together. One possibility is to make simple measurements with a tape or rule. Another is to use a *carpenter's spirit level* (Fig. 3-33). A small *torpedo* level is fine for short runs, and a 2-foot, 3-vial level is about the best for general all-around use. The 4-foot or 6-foot carpenter's level will give the most accurate indication over a long span. By using any of these instruments you can determine whether a line is dead level or whether it is off a little bit. By reading the bubbles in the vials you can also pitch a line a predetermined amount.

Another type, the *protractor* level, reads out directly in pitch. A *taut line*, which is merely a piece of heavy string stretched tight, can also be used to pitch a pipeline over a long run, or to assess a

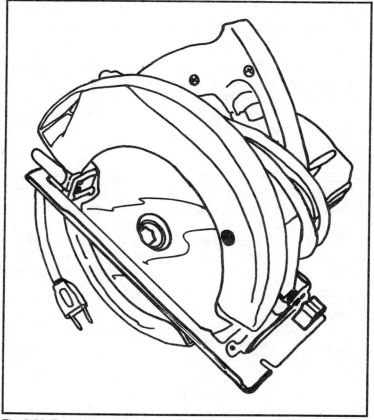

Fig. 3-31. Typical portable electric saw.

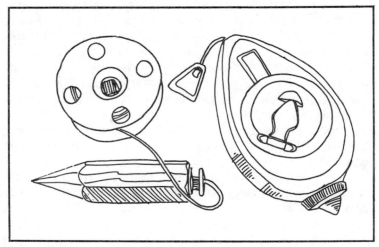

Fig. 3-32. Left, plumb bob and reeled line. Right, chalk line and chalk box that can double as plumb bob.

pipeline to see if there are any sags or dips. A *carpenter's chalk line* might also come in handy for laying out either a level line or a series of points along which a pipeline must travel through holes, notches or by direct mounting (Fig. 3-32). The chalk line will give you a direct visible guideline to follow that will be perfectly straight.

Hammers

Probably there is no possible way to get through an entire plumbing installation without recourse to a *hammer* for one reason or another (Fig. 3-34). The *carpenter's nail* hammer or *claw* hammer is the choice here, and the specific type that might be used is mostly a matter of personal preference. There are a lot of different head and claw configurations and a number of handle styles, materials and grips. But they all do about the same job with equal

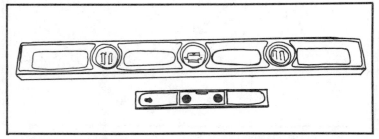

Fig. 3-33. Two-foot carpenter's level, 9-inch torpedo level.

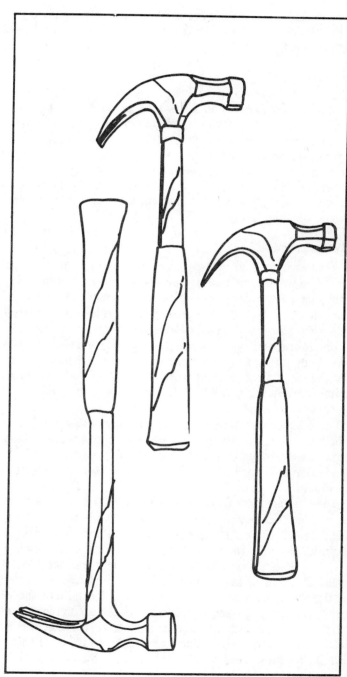

Fig. 3-34. Carpenter's 16-ounce claw hammer, center, is the most utilitarian, but 13-ounce size (below) is easiest to use on smaller jobs. Framing hammer (top) is invaluable for preremodeling demolition and structural framing.

efficiency. The 16-ounce head size is the most common and the most popular, but the 13-ounce size is a lot handier and easier to use for light-to-medium work.

Chisels

For occasional notching or trimming purposes, you might find a *wood chisel* easier to use than a saw (Fig. 3-35). For all-purpose work the plastic-handled type with metal striking cap serves well. The blade should be of hardened and tempered high-carbon steel with two precision-ground cutting edges. Such chisels come in several sizes from ¼-inch to 2-inch width in ¼-inch increments. However, you'll probably find that a couple of chisels toward the lower end of the range—say, a ½-inch and a 1-inch—will take care of your needs quite adequately.

Safety Equipment

When making a complete plumbing installation, or for that matter when working with tools in any fashion, it's an excellent idea to avail yourself of the appropriate safety equipment that is available, and of course to follow the general safety rules. In particular, working with power tools requires a degree of caution, and all such tools should be treated with great respect. Keep a good pair of *safety goggles* handy at all times, and be sure to wear them whenever there is any chance at all of sawdust, splinters, masonry dust or metal shards flying about (Fig. 3-36). A *face shield* can also be effective, but less so than goggles.

For some chores, a pair of heavy leather *gauntlet gloves* may also be in order. A quantity of fine dust in the air means that putting on a good *dust mask* or *respirator* is a smart move, particularly if you happen to have allergies (Fig. 3-37). Be sure to keep clean filters in place always.

There is one more piece of equipment that deserves serious attention, too. Considerable evidence points to the fact that sound levels above 70 dBA can be injurious. Levels of 75 dBA, equivalent to the noise generated by heavy traffic, is detrimental. By comparison, nearly all shop power tools, even the small ones, run at levels over 90 dBA. A chain saw may run close to 115 dBA. This means, then, that the smart power tool operator will protect his ears and his hearing. There is an extensive array of devices now on the market for that purpose, and all of them are inexpensive and effective. At the least, wear earplugs that will protect you up to 105 or 110 dBA. Better yet, wear a set of hearing protectors of the type that look

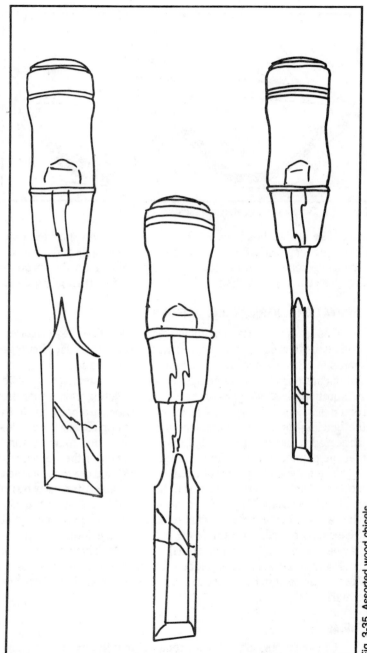

Fig. 3-35. Assorted wood chisels.

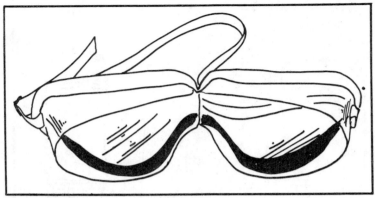

Fig. 3-36. Typical plastic safety goggles.

much like stereo headphones (Fig. 3-38). Though the idea may seem silly, especially for those of us who have happily and unknowingly gone about our noisy business for years with no such protective devices, wearing them is a whole lot better than going deaf.

RENOVATIONS, REPAIRS AND ADD-ONS

One feature of plastic plumbing materials is that they can often be used to effect relatively easy repairs in existing metallic plumbing systems. They are equally useful for making renovations, to replace and update portions of an antiquated metallic system, or to make substitutions and additions in parts of an existing system during remodeling jobs. Of course, when new additions are being made to the building, an add-on plumbing subsystem or extension can be fabricated entirely of plastic and then readily joined to an existing metallic plumbing system in the original portion of the building.

The tools and equipment entailed in this kind of work includes most or all of those already covered. In some cases there may be no need for any additional gear; this is dependent largely upon exactly what work is being done, to what extent, and the specific nature of the existing metallic system. In other cases some specialized items may be needed, and it is suggested that if you don't have the tools on hand, you borrow or rent as needed. There may also be some tasks, like pipe threading, that you can have done by an outside contractor or service shop.

Cutting

Obviously this kind of work requires that the existing metallic system be cut into in order to make the repairs or additions.

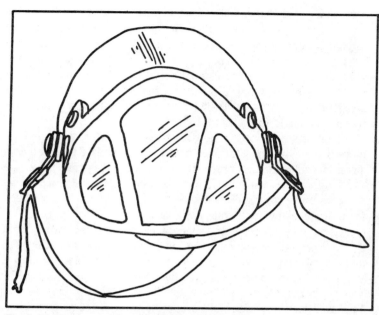

Fig. 3-37. Dust mask with replaceable filter.

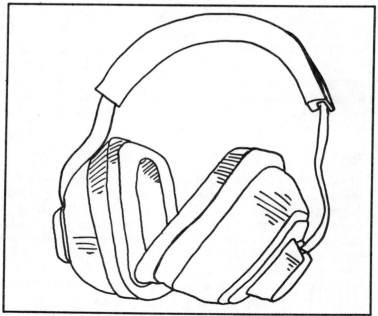

Fig. 3-38. Head-set type of hearing protectors.

Sometimes one or two cuts will be sufficient, while in other cases many more may be necessary, along with a certain amount of juggling around of the original piping in order to make things come out right. Copper tubing and copper pipe of the smaller diameters can be easily cut with a tubing cutter, as long as there is sufficient space available to swing the cutter around the pipe. Use a cutter designed for working with copper, however, and not the type specifically made for cutting plastic.

Iron or steel pipe can be cut with a *pipe cutter*, which is little more than an overgrown tubing cutter. The standard type requires a considerable amount of room to navigate because of its large size and long handle. However, there is a special variety made that employs three cutting discs and needs only to be pushed back and forth in a shallow arc to make the cut.

If neither of these tools are available or they won't work because of space limitations, you can always resort to a hacksaw or some similar type of saw fitted with a metal-cutting blade. A power reciprocating saw also works quite well for this purpose. This is the way the larger diameters of pipe must be cut, too. For pipe that has thick wall sections, use a blade of approximately 16 teeth per inch. A blade of 24 or so teeth per inch works better on thin-wall pipe or tubing.

Sometimes disassembly of a certain portion of a metallic system is possible, rather than cutting. Sometimes, too, a certain amount of reassembly in a somewhat different configuration is indicated in order to achieve the piping layout that you want. If the system is made up of copper tubing and flare fittings or compression fittings, all you need for disassembly is an appropriate wrench or two. Reassembly is done with the same wrenches, a tubing cutter and perhaps a flaring tool.

Torch

If the sytem is made up of copper pipe and brass fittings, they will be soldered together. Disassembly requires the use of a torch. Almost any kind can be used, but the most convenient is the small propane variety that uses a burner connected directly to a hand-held fuel tank (Fig. 3-39). These are inexpensive and available at any hardware store, and work quite nicely even with large copper DWV pipe and fittings. By heating the fitting and pipe until the solder melts, the pieces can be pulled apart. Reassembly is done in the same way, using fresh solder and soldering flux.

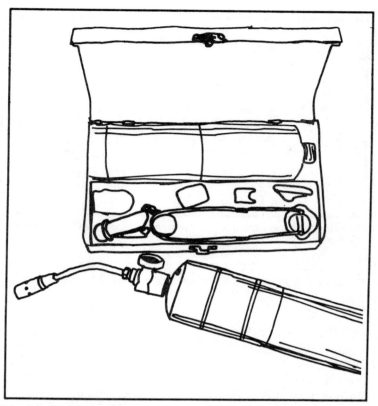

Fig. 3-39.Typical propane torch kit.

Stillson and Pipe Wrenches

Steel and iron pipes are threaded together in continuous runs. Generally one particular section cannot be removed without disassembling everything all the way back from the head of the line, unless there happens to be a union fitting somewhere in the line. Disassembly of this kind of system requires the use of *Stillson* or *pipe* wrenches (Fig. 3-40). They are used in pairs, one opposing the other. The first wrench keeps the fitting from moving, while the second unthreads the pipe section. Stillson wrenches are available in a number of sizes, as regards both handle length and jaw capacity. The small ones often are not terribly effective in dismantling an old system that is well rusted together. The best bet is a pair of 18-inch or 24-inch wrenches, so that you can get maximum leverage. It is also quite possible to use a Stillson in combination with the chain or reversing pipe wrench mentioned earlier, or to use a pair of such

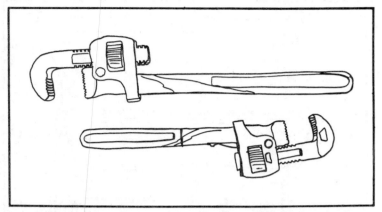

Fig. 3-40. Venerable Stillson or pipe wrenches, indispensable for many plumbing chores.

wrenches instead of pipe wrenches. Usually, however, the available leverage and grip is not quite as good with a chain wrench.

Threading

Since steel and iron pipes are threaded wherever they join a fitting, some means must be at hand to form threads in the cut ends of the pipe sections. (The fittings, of course, are factory-threaded). This can be done manually with a *pipe threader stock* outfitted with a suitable size of *pipe die* and *guide bushing*. Either two-handled regular-action threaders or single-pawl ratchet-action threaders are available for pipe diameters ranging from ¼-inch nominal to 2-inch nominal.

Threading pipe by hand is indeed a lot of work, and a power pipe threader makes the task a great deal easier. However, these rigs are very expensive and not worthwhile for the home shop. About the best answer for the do-it-yourself plumber who needs only a few threads cut is to transport the pipe sections, or the measurements of the pieces he needs, to a contractor, lumberyard or plumbing supply house and have them do the work. The charge will be more than reasonable in view of the labor saved.

Dressing Tools

Cutting metallic pipe, whether with a hack saw or a pipe or tubing cutter, gives rise to one of the same problems that occurs in cutting plastic pipe. Cutters leave a deep and sharp ridge of metal, but instead of being on the outside of the pipe as with plastic, the

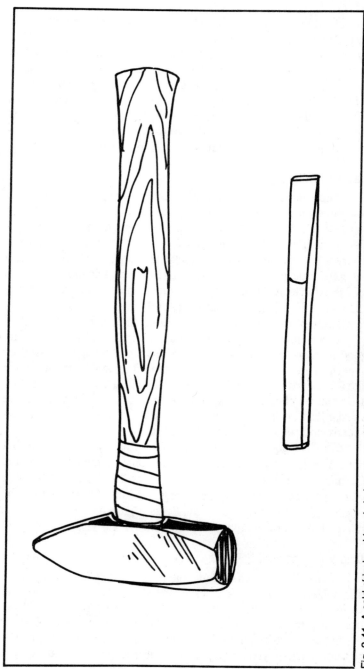

Fig. 3-41. A cold chisel and hand sledge are good for chipping out concrete and other heavy work.

biggest ridge is on the inside of the cut. Saw cuts leave burrs and ragged edges. This can be taken care of with a good file, and tubing cutters are usually fitted out with reamer blades. Pipe cutters, however, are not and so an individual reamer made for trimming off the inside edge of the pipe is necessary. There are several varieties of these. One type fits into a carpenter's bitstock, while another has its own ratcheting handle. The more expensive types probably will not be necessary, and in fact you can use any method that satisfactorily does away with the burrs and sharp, curled inside edges.

Masonry Cutting

When doing remodeling work or making additions to a system, sometimes it's necessary to run through a poured concrete or concrete block wall. Installing a sump pump or a small sewage ejector tank may necessitate taking up a section of concrete floor. There are a number of ways to tackle this problem. One is the old tried-and-true method that makes use of a combination of a large *cold chisel*, a heavy *hand sledge* and liberal amounts of elbow grease and sweat (Fig. 3-41). You can substitute any one of several different kinds of *rock* or *star drills* for the cold chisel.

Using power equipment eases the strain considerably. One good method is to employ a *rotary hammer drill* or *rotary impact wrench*. These tools are essentially the same, and look very much like a standard ⅜-inch or ½-inch electric drill. The difference is that a certain number of heavy impacts or jolts per minute bang the special drill bit into the masonry as the bit turns. The result is speedy and relatively easy hole drilling.

Another good choice is a small *electric impact hammer* that holds various kinds of chisel bits. The constant heavy vibration of the hammer drives the bit into the masonry. For big jobs where the going is tough and where a fairly large amount of material must be removed, you can opt for a full-fledged *jackhammer*, which will make short work of any such project.

Cast-Iron Pipework

Working with cast-iron soil pipe requires a good bit of specialized equipment as well as some skill and knowledge in order to do the work successfully. Lead and a *plumber's lead pot, caulking irons, oakum, yarning tools* and the like are needed. If you have to break into a cast-iron drainage system in order to make a connection to a new section, you might prefer to call in an expert to take care of the job.

An alternative to all the specialized gear, if it is allowable in your area, is to remove part or all of one section of soil pipe and make a replacement with the necessary smaller replacement pieces set together with lead wool driven in place with a caulking iron and hammer. The unwanted section can be removed by cutting it with a *soil-pipe cutter* or *pipe cracker*, or by cutting a groove all the way around the pipe with a hacksaw and then tapping it with a hammer and cold chisel.

SUPPLIES

As you go about the business of putting your plumbing system together, there are certain supplies that you will want to keep on hand all the time. Some of them are perfectly common and you will likely already have them. Others are applicable to particular aspects of the plumbing job itself.

Working with some kinds of plastic pipe and fittings, like PVC or CVPC, requires the use of two fluids in order to make the proper joints (Fig. 3-42). One is a *cleaning fluid* that removes any grease and dirt from the surfaces to be joined. The other is a *solvent-welding* liquid that temporarily softens the plastic and allows the pieces to meld together and fuse into a solid unit.

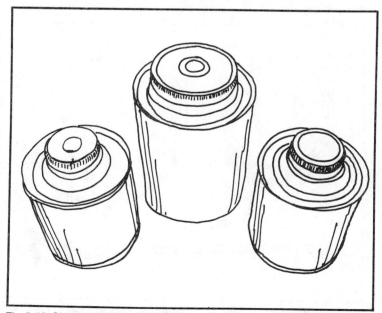

Fig. 3-42. Cleaners and welding solvents used with certain kinds of plastic pipe.

Copper pipe or tubing also has to be clean at the tip before sweat-soldering to a fitting. Fine steel wool does an excellent job of cleaning and polishing, and fine sandpaper will do an equally thorough job but is a bit more troublesome to use. Fine sandpaper is also sometimes handy for deburring plastic pipe.

Though pipe runs are sometimes set in notches or bored holes in the structural framework of a building, they also are frequently suspended from the framework and held in place by pipe clamps. Special plastic clamps are made to hold the smaller sizes of plastic pipe, and they are secured with small nails. Clamps or hangers can also be bought for large-diameter pipe, but they are quite expensive. A frequently used substitute is perforated metal strapping, which can be bought at hardware stores or plumbing supply houses in small, inexpensive rolls. Since the perforations are rather large, broad-headed roofing nails work about the best for securing the strapping.

Soldering copper pipe and brass fitting combinations obviously means that you will need a roll of *solder*. Under no circumstances should an acid-core solder be used and the rosin-core solder commonly used in electronic work is not the proper kind, either. Instead, use a straight, uncored solder of the kind normally used for plumbing work. A ⅛-inch diameter wire form of solder works about the best, and is available in convenient ½-pound and 1-pound rolls or

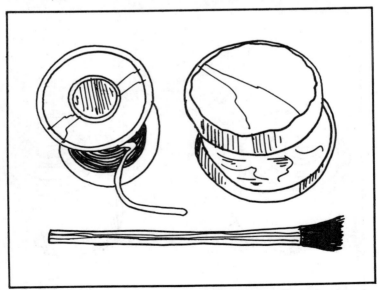

Fig. 3-43. Roll of wire solder, paste-type flux and flux applicator brush.

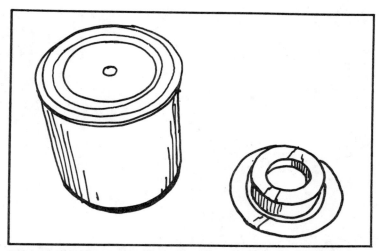

Fig. 3-44. Pipe joint compound and plumber's Teflon pipe joint tape serve the same purpose—sealing pipe joints.

spools (Fig. 3-43). As to its composition, 50-50 is fine (50 percent tin, 50 percent lead), and 40-60 (40 percent tin, 60 percent lead) is also perfectly acceptable, though a bit softer.

Since this type of solder has no *flux* contained in an integral core, the flux must be applied separately. A small tin of noncorrosive paste flux (Fig. 3-43), which is actually a rosin-base type of flux, will see you through a great many pipe joints. The paste can be applied to the surfaces being joined by wiping it on with a bit of cloth, or your finger, or by applying with a small metal-handled brush (Fig. 3-43) sometimes referred to as an *acid* brush. These are usually available wherever solder and flux are sold.

Joining the threaded connections in iron and steel pipes and fittings is never done dry. If they are so joined, in the first place they will not drive together properly, and in the second, they'll leak like a sieve. There are two choices of material to use on the pipe threads. The oldest and the best-known is *pipe-joint compound*, or *pipe dope* as it usually called (Fig. 3-44). The dope is always applied to the male threads, and lubricates the threads as they are joined. It helps to prevent rust and seals the threads up tight. The second choice is *Teflon plumber's tape* a much newer but equally effective material (Fig. 3-44). This actually is a tape that comes in a small roll, and a couple of turns around the male threads does the job. Though more expensive, the tape makes a good deal neater job and is a lot less messy to work with.

Chapter 4

Working With Plastic Materials

As with any other area of endeavor, the key to assembling a successful and effective, long-lived plastic plumbing system, or any subsystem or section thereof, lies in knowing not only what materials to use, but also how to use them. There are various practices, procedures and techniques that come into play, none of them difficult nor requiring any great degree of skill, but only a bit of knowledge, common sense and forethought. Once you understand a few tricks of the trade, along with some general hints and kinks, you should have no trouble whatsoever in single-handedly putting together a fine plastic plumbing system. Here, then, are a number of areas of general concern that require special consideration and are pertinent in many aspects of assembling different kinds of plumbing subsystems. All of them are aspects that can present problems and pitfalls for the unknowledgeable or careless workman, and all of them have led at one time or another, when disregarded or improperly accomplished, to plumbing system difficulties or failure.

DESIGN AND WORKMANSHIP

Any discussion of the fabrication of a plastic plumbing system, or any other kind, for that matter, presupposes that the overall design of the system or subsystems is correct and properly done, and that the workmanship in assembling the system is of high quality. More will be said about designing plumbing systems in a later chapter, but suffice to say for the moment that if the design of the system is incorrect or faulty, the system will probably not go together easily or correctly. Also, it will not function effectively when the job is done. Inevitably corrections and rework will have to be undertaken to get the system into suitably operational condition. The system must be designed not only in compliance with local or national plumbing codes, but also in accordance with good, tried and proven general plumbing practices. The system must also be designed to comply with certain laws of physics and mechanics. Water not under pressure will not run uphill, unsupported or improperly laid pipes will droop and sag, drainage flow will not take place correctly without proper venting, and so on.

It is also true that unless a system is assembled in a workman-like manner, using the proper techniques, procedures and practices, difficulties will arise somewhere along the line. This may happen during the assembly itself, or may not show up for months or even years in the form of leaks, stoppages, breakage or other malfunctions somewhere in the system. Trying to define good workmanship is difficult, but that makes the quality no less important. In essence, this is a matter of selecting the proper materials and supplies, assembling them with the correct tools and procedures, and working in a careful and neat manner. Patience is important, and so is careful thought and planning. A willingness to do something over again if the task does not come out quite right is just as important. Thoroughness in doing each individual task, followed by careful inspecting and/or testing, can forestall a good many problems. It has often been said that there are two ways to accomplish a chore, right and wrong. This is particularly true of plumbing systems. About 90 percent of the plumbing system in your house, once completed, is buried deep within the house structure or underground and inaccessible. Do the job the right way, and you should never have to start ripping and tearing to find out what went wrong.

CHOOSING PROPER PARTS

Choosing the proper plumbing parts for each individual task at hand is an extremely important matter. The best way to come up with the right answers is to study the situation in some depth as it relates to each particular plumbing problem that you run up against. Manufacturers' catalogs, which can be obtained direct or examined at plumbing supply houses, furnish a substantial education in plumbing system conponents. Here you will find an incredible array of pipes, fittings and accessory gear laid out for you, often with capsulized information on what each particular part is and how it is used. Conversations with plumbing supply house countermen are also very productive, especially when you run into a particularly sticky problem for which you can't find an answer. These fellows, many of whom have had years of experience in the plumbing trade or plumbing supply business, have tremendous stores of information tucked away in their heads and seemingly are able to solve practically any plumbing problem. In fact, they have seen most problems a number of times before. Another source of information, of course, is books like this one, from which you can derive considerable amounts of knowledge and instruction.

The reason that choosing the correct parts is so important is simply that there is a vast range of products made for plumbing purposes, many of them especially designed to be properly functional only under certain circumstances. If you use the wrong material, they may work inefficiently, poorly or perhaps not at all. To take an example, certain types of plastic pipe are designed to carry potable water for drinking and other domestic purposes. Other types, though perhaps made of the same base materials, are not. If you use an unapproved type of pipe for water supply purposes, you could endanger your health.

Another important aspect lies in the choosing of correct fittings. You'll notice, for instance, that water supply fittings commonly make abrupt, right-angle turns. The water is under pressure and is forced through these sharp bends with no difficulties and no great loss of system efficiency. On the other hand, fittings used in pipelines that carry human waste always make gentle bends. The liquid is not under pressure and must flow against the least possible resistance, and for the most part it makes few if any sharp turns. Also, certain drainage fittings, as in a sanitary tee where a horizontal branch line connects to a vertical main line, are designed with a special rounded corner that must always be placed at the lower side, so that the waste can flow smoothly into the stack with no hangups. By comparison, a bullnose tee that is used primarily in horizontal drainage fields or leaching fields does not have this special curved inside edge, because flow characteristics here are of little consequence.

In short, if you are not sure as to what kind of pipe material might be the best for some particular use, or what kind of fitting might best be employed in a particular situation, investigate first before going ahead.

Laying pipe is a process of uncoiling flexible pipe or joining sections of rigid pipe and arranging the pipeline in a trench as an underground run. This is an area where problems frequently arise because of improper handling of the materials. But if correctly done, there should never be any difficulties nor any need to dig the line up again for repairs.

LAYING PIPE

When installing a water supply line of flexible plastic piping, try to keep below-ground couplings to an absolute minimum. This is not only saves fitting costs and labor, but also makes a better installation simply because every joint or connection can be a potential source of trouble, a weak point (comparatively speaking) in the line. Connections made to a saddle at a water main should include an expansion loop (this is sometimes made from a short piece of Type K copper), with the tap being made headed upward at about a 45-degree angle and the pipe curving back down in a question-mark shape to the trench floor. Then, should the main or the new supply line shift, or the loose backfill settle down and move the line, the pipeline has a bit of slack and can move without pulling the connection loose from the main. The same arrangement is used with rigid supply lines.

The supply line should be carefully laid on the trench bottom. For flexible pipe, the line should be uncoiled in the opposite direction from the original coiling, by securing the pipe at one end and unrolling the coil backward until it lies flat. This avoids kinking and twisting and unwarranted strains on the pipe walls. Snaking the pipeline back and forth on the trench floor allows plenty of expansion-contraction movement. Rigid sections should be fitted together with extra care, with all couplings carefully installed and checked. In both cases, the trench floor should be smoothed with a square-bladed shovel or a steel rake, and all rocks or protrusions removed from beneath the pipe. Level and/or grade changes should be made in gentle arcs, never in abrupt steps, and the pipe should be fully supported by earth throughout the entire run.

It is wise to make all connections—tap at the main, supply connection at the house, pressure reducing valve, outside hydrant or whatever—throughout the entire pipeline before starting any of the backfill work. Just as soon as connections are made, turn the main valve on and allow the pipeline to fill under pressure. Keep the pressure on for about four hours or so, and thoroughly check every joint in the line for leakage or seepage. Only when the line has been found to be absolutely tight and leak-free should the backfill work proceed.

Fig. 4-1. Special plastic pipe straps secure CPVC water pipe firmly, but allow expansion-contraction slippage without noise.

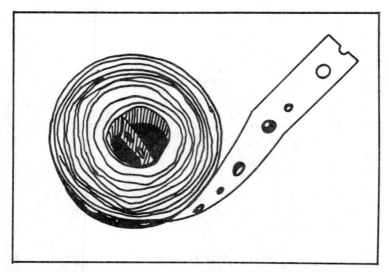

Fig. 4-2. Perforated pipe strap can be cut off stock roll in whatever length is needed for use in securing or suspending pipelines.

The initial backfilling should be done with considerable caution. The first step, regardless of the type of pipe in use, is to bury the line to a depth of about 4 to 6 inches with clean, loose fill. Do this job entirely by hand, making sure that there are no rocks larger than tennis-ball size in the fill. Power machinery can then take over the job, but even then the dumping of large rocks down into the trench is to be avoided. Save exceptionally rocky fill for the uppermost part of the trench. If the backfill consists mostly of rock, as is sometimes the case, some clean fill such as sand, rock-free subsoil or even gravel should be imported to bed the pipeline to a depth of about a foot. The trench can then be carefully topped up with the rocky spoil dirt. If the backfill is mounded 6 to 8 inches or so higher than the surrounding grade level along the trench line, eventually it will settle out to an approximate level.

Waterlines that are buried only a foot or so deep, such as supply lines in perennially warm climes or sprinkling systems, are handled in much the same way. Here the backfilling process is less troublesome, but nonetheless the pipeline should be protected from settling or being crushed by heavy rocks. Also, in the interest of being able to completely drain the lines if necessary, they should be laid flat, with no droops and sags. Pipelines close to the surface should be deep enough so that any mechanical damage, as from a heavy truck driving over them, is unlikely to occur.

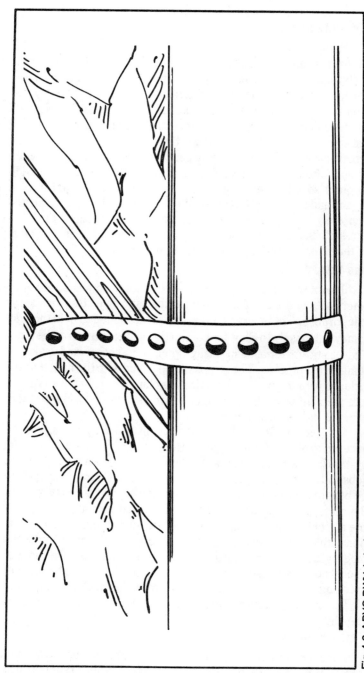

Fig. 4-3. A PVC-DWV drainpipe suspended from joist with perforated pipe strap.

149

HANGING PIPES

Hanging pipe would seem to be a pretty simple affair, and indeed it is, provided that you know how to do it. All kinds of rigid plastic pipe must be given adequate support; even though rigid, the pipe has a certain amount of limberness and should not be counted upon to support itself. Vertical runs of DWV and similar pipe will stand straight and remain so without additional support provided that the total run is short, on the order of a section and a half or so depending upon pipe size. But longer runs should be braced in some fashion (often structural framing members of the house will provide support) so that they cannot move around, but are still free to expand and contract. Horizontal or sloping lines are best supported every few feet, with the specific spacing dependent upon the inherent stiffness of the pipe sections. Supports should be close enough so that there is no sag from support to support.

Pipe Straps

Plastic water supply lines can be secured to joists, studs or other parts of the building by means of plastic pipe straps made for that purpose. These straps are especially designed to hold the pipe somewhat loosely to allow for contraction and expansion of the pipe. The plastic materials slip easily upon one another and make no sound in the process. The straps can be put up with the nails that are included with them, but the meticulous plumber might like to substitute suitable wood or sheet metal screws (Fig. 4-1). Space them every 2 ½ feet maximum.

Larger piping, such as that used in DWV systems, can be secured in place with metal pipe straps of the type used on metallic systems; however, these are relatively expensive. An adequate substitute is the perforated soft galvanized metal strapping that can be bought in rolls at hardware or plumbing supply stores (Fig. 4-2). Supports are simply arranged by cutting off an appropriate length of the strapping with tinsnips, securing one end through the perforation with a broad-headed roofing nail, wrapping the strap under the pipe and back up, and nailing the second end (Fig. 4-3). The pipeline should not be snugged tight against the mounting surface wherever creaking noises from expansion and contraction might be objectional. This same system can be used where pipes must be suspended in the air.

Types of Support

Where drainlines or similar piping must be run across an open space, as in a crawl space, for instance, other types of support may

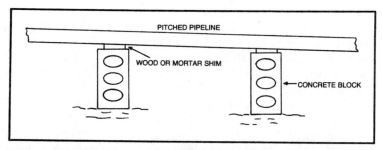

Fig. 4-4. Horizontal drainpipe solidly supported above ground in crawl space or other area by setting upon concrete blocks and wedging/shimming as necessary to maintain pitch.

have to be arranged. This can be done with special pipe hangers that clamp around the pipe and are joined to a flange mounted overhead (and any practical distance away) by a length of threaded rod. An easier and less expensive method may be to support the pipeline from below on flat stones or concrete blocks, with scraps of wood serving as shims for the final leveling (Fig. 4-4).

For a cleaner and more workmanlike installation, leveled posts of redwood or other treated wood can be set in the ground with the pipeline laid atop them. Where a continuous accurate slope is needed for proper drainage, a grade board of 1 × 6s or 2 × 6s can be attached to the posts, or to stakes driven in the ground. The pipeline is then continuously supported and given no opportunity to sag. Also, the pipe can be secured to the posts or grade board with perforated metal strapping so that it cannot shift out of line.

Framing Members

Pipes can also be run through holes bored in joists or other framing members. If this method is used, as it frequently is for water supply lines, it is important that the holes line up accurately with one another. The easiest way to get a straight-line run is to snap a chalk line across the lower edges of the hoists as a guide for drilling. Each hole should be centered the same distance above the bottom edge of the framing member, preferably at least 2 inches. The hole should be bored straight through the member, not cocked off at one angle or another. This is more easily done with a right-angle drill or drill attachment that will fit between the standard framing member spacing of 16 inches on centers. All of the splinters should be removed from around the bored hole, which should be about ¼-inch larger than the outside diameter of the pipe being used.

Another method of mounting is to cut small notches in the edges of framing members and set the pipeline into them (Fig. 4-5). Each notch should be about ¼-inch deeper and ¼-inch higher than the outside diameter of the pipeline. As with the bored holes, the notches should all line up. Cutting the notches is easily done by setting the shoe of a power circular saw to the desired cutting depth and making two passes across the face of the member the desired distance apart. A sharp rap with a hammer will pop the small block of scrap wood free. Once the pipeline is set in place, each notch should be closed with a piece of metal strap secured to the wood. For most lightweight water supply pipeline installations, a short piece of perforated metal strap is adequate. Depending upon the nature of the finish covering to be later applied over the framing members, these straps may have to be recessed slightly into the framing member so that no bump or ripple will appear on the finished surface.

Pipes larger than nominal ¾-inch size can be notched into framing members or run through bored holes, but this is a practice to be avoided wherever possible. The reason is that large notches or holes in the framing members reduce or may even completely negate the strength of the member at that particular point. The framing is thereby weakened to a certain degree. Furthermore, running large pipe in this manner is a lot of extra work, provided that another method can be found. The common exception to this is in the running of vertical stacks, which usually pass through the sole and top plates of wall frames. Loss of strength is not as critical in this situation, but even so additional beefing up at these points with metal straps or additional small framing members is a good idea.

PIPELINE PLACEMENT

Frequently, especially in residential applications, the design and subsequent plans for a plumbing system are done schematically or symbolically, with primary regard given to the necessary pipe, fittings, accessories and fixtures, but not the specific routing or placement of the lines themselves. Many plumbing jobs are done without benefit of any plans at all, but simply a list of required fixtures and their locations. In both cases, it is left up to the plumber to choose the best ways of getting from point to point with his pipelines. This is all well and good if the plumber knows his business—and is also aware of the overall design and plan for the building, so that his work does not interfere with other parts of the project. However, sometimes peculiar things happen.

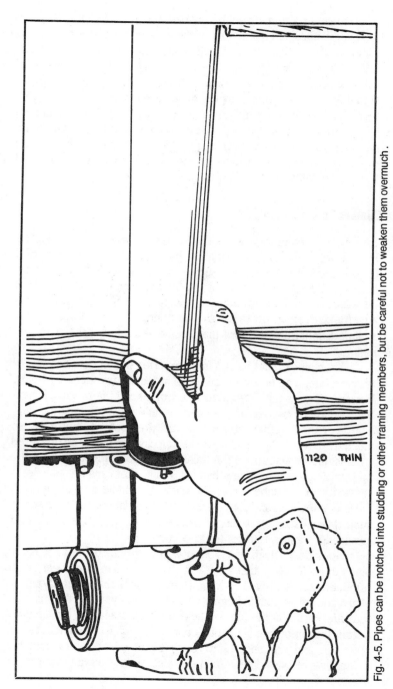

Fig. 4-5. Pipes can be notched into studding or other framing members, but be careful not to weaken them overmuch.

Pipeline routing should be done with care and a knowledge of the overall building layout, so that the lines will not later have to be changed or rearranged because they interfere with something else. It does happen that pipelines get run smack across the middle of window or door openings, pass through a fireplace footing or get somehow tangled up in complex structural framework. The shortest and straightest route from point to point with a minimum number of fittings is always the best from the standpoint of both the plumber and an efficient plumbing system, bit it may not be so from other angles. Choose each pipeline route with care in order to avoid future problems.

Climate Considerations

Another common error is that of running water supply pipeline in exterior walls in cold climates. The excellent possibility of a freezeup somewhere along the line is simply disregarded. Sometimes the lines are even installed on the exterior side of the thermal installation, which means that they almost surely will freeze eventually. Make every effort to run all water supply lines through interior wall partitions, under floors that lie above heated areas or in ceilings.

When pipes must be run in an exterior wall in climates where temperatures can be expected to fall at least a few degrees below freezing, always run them to the interior side of the thermal insulation and vapor barrier. If they can be hidden within closets, cupboards or the like that remain essentially at room temperature, so much the better. Pipelines run in ceiling cavities below an unheated attic should lie just above the finish ceiling covering and below the thermal insulation and vapor barrier. Piping that runs through flooring cavities over unheated crawl spaces should be run just beneath the floor sheathing and above the thermal insulation, but even so it is subject to possible freezeup if temperatures fall low enough or if localized air infiltration or drafts occur.

No matter which way you look at it, the best bet is to route pipes in such a way that they will always remain at or near room temperature. It is well to keep in mind, too, that even though a structural cavity that is bounded on one side by a cold wall may be tight and secure at the time of the plumbing installation, buildings do settle. Insulation can shift, tiny cracks through to the outdoors can open up, and the once-tight cavity becomes no longer suitably protective of waterlines. In the event of cold air infiltration, as during a severe winter storm, for instance, freeze up is almost a sure thing.

Drainage Line Routing

Drainage lines are not quite as problematical as water supply lines, but nonetheless they must be correctly installed for proper operation. As far as routing is concerned, a straight course is the best course in most cases, and the pipeline should be routed so that nothing interferes with this, or vice versa. Drainage lines, and especially leaching field lines, that are not buried too deeply should be positioned so that passage over them by heavy trucks or equipment is unlikely. Make sure that later construction, such as additions to the house, a swimming pool or other similar appurtenances, won't occasion a major upheaval and disruption of the sytem.

Freeze up of drainage lines is a much more common occurrence than you might suspect. Often the cause can be traced to improper installation, such as a sag in the line, use of improper drainage fittings, insufficient slope or similar circumstances that lead to a blockage and backup in the line, with subsequent freezing. Long runs of above ground drainage lines in unheated crawl spaces, especially if run at a gentle slope, can also lead to freezeup. The simple solution to the problem is to install the runs correctly and get them below ground and out of unheated areas in the shortest possible exposed pipeline distance. In such circumstances, cold winter wind is perhaps the greatest enemy of exposed drainage lines. Route them away from foundation walls or skirting where cold air jets blasting through cracks cannot reach them. Wrapping the pipes with insulation is a waste of time for the most part, since no heat is generated within the pipe for the insulation to retain. Pipeline, insulation and surrounding air remain at equal temperatures most of the time.

RUNNING TUBING

Running plastic tubing, such as the relatively new PB, is a simple task, much easier than installing other kinds of pipe. And though a very forgiving material, there are a couple of pointers that will help you in the installation procedure.

Because this material is so flexible, it cannot be run in neat, straight lines like CPVC. It droops and sags and dangles, and cannot be run to a constant pitch, at least from a practical standpoint. Therefore, if your plumbing plans call for a water supply system that must be fully drained periodically, as in a summer home that is closed up and unheated during the winter months, this may not be the pipe for you. Flexible tubing cannot be satisfactorily drained, though if the system is opened and the pressure relieved, the tubing will simply expand when water freezes within it, with no damage. Water trapped and frozen in a fitting, though, will rupture it. An alternative would be to arrange the system so that compressed air can be blown through it to clear the lines.

Tubing can be run through small bored holes in structural members. If supported in this manner, care should be taken when the long lengths of tubing are pulled through the holes and into place, so that the tubing does not kink or snarl, or catch and tear on a rough edge. As long as it is carefully fed into place, no damage will result. Often as not, however, flexible tubing is run or pulled through cavities much after the manner of Type NM electrical cable. This is a most satisfactory method of insulation, since it is quick and simple. Again, the tubing must be pulled into place with care so that it does not snag, kink in half or become otherwise damaged. Usually the tubing can be left to lie wherever it happens to be, but if supporting a section of tubing is necessary, a wrap of copper electrical wire loosely coiled around it and attached to a nail is as effective as anything else. Small-sized tubing can be held in place with large electrical staples, provided that the staples are not driven down tight and plenty of clearance is left all around the tubing as it passes through the staples.

INSERTION FITTINGS

Insertion fittings made of either nylon or metal are commonly used with flexible polyethylene water supply and irrigation piping. Each end of plastic-to-plastic fittings is made with a series of sharp-edged molded-in rings, the diameter of which is just slightly greater than the inside diameter of the pipe. Putting these fittings together is simple enough, but requires some effort and often the fittings are balky.

The procedure is to first slip a pair of loosened stainless steel clamps down over the pipeline and back out of the way. Then push the fitting into the pipe. It helps to hold the pipe firmly in one hand and the fittings in the other. Wiggle the two back and forth in opposing directions (Fig. 4-6). The pipe walls must be forcibly expanded to a small degree as the fitting is pushed in, and it won't go easily. However, persist until the fitting is driven all the way home to the fitting shoulder and can go no farther. Under no cir-

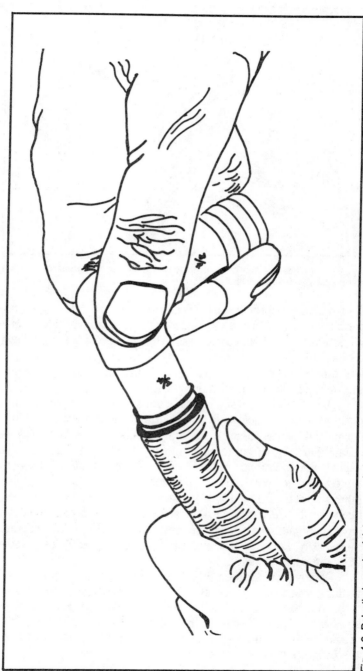

Fig. 4-6. Polyethylene pipe joints are made by shoving the insertion fitting ends directly into the pipe.

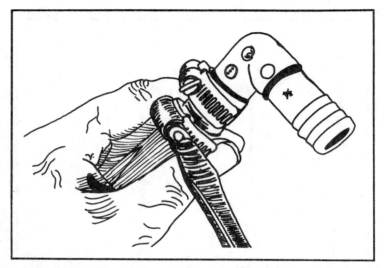

Fig. 4-7. Tightening a pair of worm-drive stainless steel clamps around each fitting end completes the joint in PE pipe.

cumstances should an insertion fitting in a pressure line be left only partially inserted. Nor is it a good idea to use soap, vaseline or any other lubricant to help the process. The whole idea is for the rings on the fitting to bind firmly against the inside pipe wall; though a lubricant makes it easier to put these fittings in, it also means that the fitting can come apart more easily.

Once the fitting is seated, slide the stainless steel clamps into position, one just below the shoulder and the other a bit further down, and tighten them as much as possible with a large screwdriver. Don't be afraid to apply plenty of pressure, because it is unlikely that the clamp will break. You will not cause any damage to the pipe. To complete the fitting joint, repeat the process at the other side of the fitting (Fig. 4-7). Incidentally, no pipe dope is really needed with these fittings, but some plumbers prefer to use it anyway. Sometimes it does help the pipe and fitting to slide together more easily, but it will not render the fitting more easily removable after the dope has set up.

Removal of insertion fittings by simply pulling them out is a virtual impossibility. The best method is to cut the pipe off an inch or so beyond the fitting end and carefully slit the pipe wall open with a sharp utility knife. If enough care is exercised, and the fitting rings are not scored by the knife, the plastic pipe can be peeled away and the fitting reused.

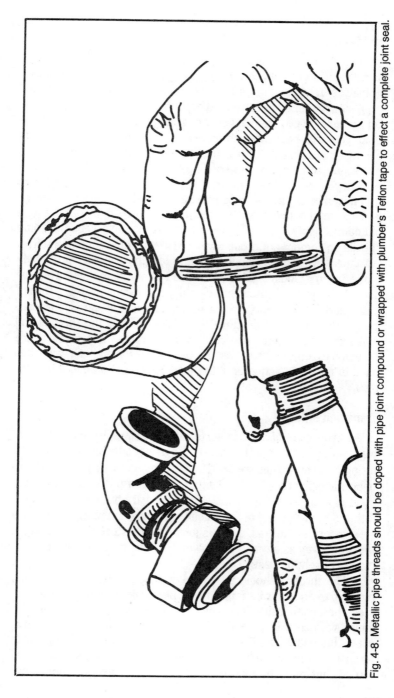

Fig. 4-8. Metallic pipe threads should be doped with pipe joint compound or wrapped with plumber's Teflon tape to effect a complete joint seal.

159

THREADED FITTINGS

Threaded fittings are most often encountered in metallic plumbing systems and especially in water supply piping. Galvanized or black steel pipe is threaded, some brass piping is threaded, and copper tubing employs threaded brass fittings of either the flare or compression type. Though you are making plastic plumbing installation, you will encounter threaded metallic fittings at numerous points, such as in adding to an old system, piping up a water heater, or making connections to various plumbing fixtures.

If you need threaded sections of pipe when working with galvanized or black steel pipe rearrangements, the easiest method is to have the necessary pieces cut and threaded for you at a plumbing supply house. This is a laborious job that requires some special equipment not ordinarily found in the home shop nowadays. An alternative is to buy nipples or short lengths of stock pipe already made up in standard sizes. Threaded joints in metallic systems should always be treated with pipe dope or wrapped with a turn of plumber's Teflon tape (which accomplishes the same purpose) before the joint is made (Fig. 4-8). This will seal the threads and make the joint leakproof. The fittings, or pipe and fittings, are screwed together with the aid of a pair of Stillson or chain wrenches used in opposing action until the joint is as tight as possible (Fig. 4-9).

Transition Fittings

Plastic pipe of various sorts can be adapted to galvanized or black steel pipelines or fittings by the use of adapters or transition fittings. These fittings are designed to connect to the particular type of plastic pipe at one end and either thread onto (female fitting) the steel pipe or thread into (male fitting) a galvanized or black steel fitting. Often this end of the adapter or transition fitting will have a steel insert and thread, and it is joined to the steel pipe or fitting in the same way as any other. However, some transition fittings are made entirely of plastic.

The threads should be doped or taped as usual, but caution must be exercised when turning the plastic fitting into (or onto) the steel threads, because it is easy to crush or mangle the fittings with a pipe wrench. Also, the threads, if plastic, will strip with relative ease. Instead of a Stillson wrench, use a strap, an adjustable wrench, or a mechanic's open-end wrench. The fitting should not be forced, but must be made up snugly. The insertion-type fittings

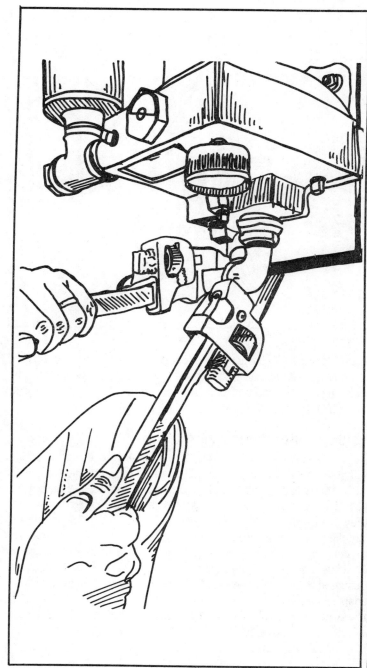

Fig. 4-9. Assembling pipe and fittings on a water pump with a pair of pipe wrenches. Note opposing jaw action.

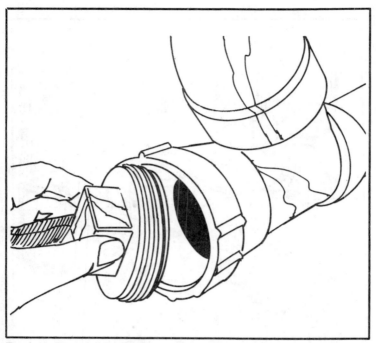

Fig. 4-10. Threaded PVC-DWV cleanout and plug.

made for transition purposes are metal. These too can be deformed or crushed, or the threads stripped, with relative ease. Make the joint carefully and do not force the threading. Again, pipe dope or tape should be used on the threaded portions.

Plastic-to-Plastic Threaded Joints

Occasionally one will find plastic-to-plastic threaded joints. These are rare, because the wall thickness of plastic pipe is usually such that threading cuts the pipe strength to shreds, and so they are not very effective. But there is one particular application where they are used, and that is in cleanouts (Fig. 4-10). Cleanouts are installed in one or more locations in the DWV system to allow access for cleaning the pipeline out with an auger in the case of obstruction. A cleanout adapter fitting is joined to one arm of a wye fitting, and is designed to accept a removable cleanout plug that threads into the adapter (some, however, lock and seal in place). These parts are made of substantial thicknesses of plastic and have sturdy, large threads for easy operation. Nonetheless, they should be treated with some degree of care.

The plugs are best not removed or replaced with a toothed wrench like a pipe wrench, but rather with a large crescent or with a strap wrench. They should be taken down snug, but not overly tight, with support given by a second strap wrench to the main body of the fitting during the initial stages of loosening the plug or the final stages of tightening it up. Pipe dope is best used on these threads, too.

Another spot where plastic-to-plastic threaded joints are found with even more frequency is in plastic trap assemblies of the adjustable type, and in trap adapters that fit the trap extension to the drainpipe line (Fig. 4-11). These fittings contain seals that actually do the work of enclosing the jointure, and do not depend upon the thread for sealing, only for mechanical coupling. These fittings should be tightened by hand to a firm, snug seat, but never force them with a wrench. There are also several kinds of transition

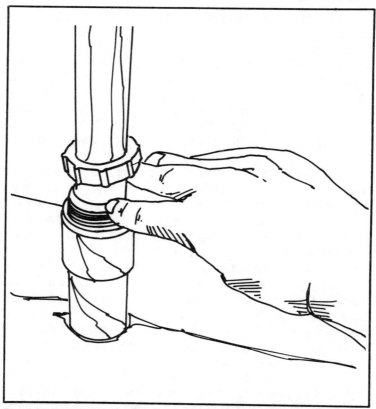

Fig. 4-11. Typical threaded slip nut and trap adapter on PVC-DWV line.

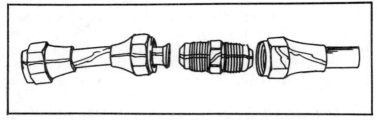

Fig. 4-12. Flare fittings on copper tubing; brass coupling joint.

fittings and adapters made in a similar fashion enclosing metal grab-rings, neoprene or nylon washer-type seals or similar devices. As with trap slip nuts, these fittings should likewise be only snugly seated and never run down wrench-tight.

FLARE FITTINGS

Flare fittings have long been used on copper tubing of all kinds and in numerous applications. They are frequently found in household plumbing systems where copper tubing is used. A flare fitting consists of two or three machined brass parts: the fitting body, and either one or two flare nuts (three in the case of a tee) depending upon the type of fitting (Fig. 4-12). The end of the copper tubing must be flared to match the chamfered internal edge of the fitting body, while the outer rim of the flare forms a flange that is clamped down tight upon the fitting body by the flare nut, making an effective seal.

Attaching a flare fitting is not a difficult job, but it must be done with care in order to insure an effective flare seat and consequent pipeline seal. The job is done with a standard flaring tool, which is available at any well-stocked hardware or automotive supply store. The process begins by first slipping the flare nut onto the tubing and down out of the way. The tubing is then clamped into the proper diameter hole of the flaring jig, with a certain amount of the tubing protruding above the jig surface. The flaring cone is clamped into place and run down into the tubing, expanding the material and stretching it out into a lip. Making up the fitting consists merely of setting the flare on the fitting body seat and turning the flare nut down onto the threads (Fig. 4-13). These fittings should be made up wrench-tight; dope is not used.

Interestingly enough, the flaring process can also be used with either CPVC pipe or PB tubing, provided that an extra step is taken. Before setting the tubing in the jig, immerse the end in a pot of boiling water and heat it up thoroughly. Immediately place the

164

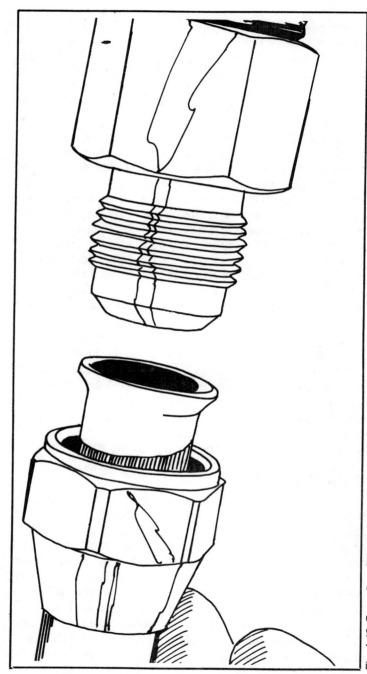

Fig. 4-13. Brass flare fittings can be used with flared CPVC or PB pipe or tubing.

tubing in the jig and make the flare (Fig. 4-14). The end of the tubing must be absolutely square and set in the jig with just the right amount of tubing protruding. The flaring cone should be run down smoothly and fairly slowly, but not so slowly that the tubing will cool off before the flare is made. If the flare flange has ragged edges or is cracked in the process of flaring, the joint will not be successful. Cut the end of the tubing off and start over again.

Although it can be done, connecting CPVC pipe with brass flare fittings is not a common circumstance because the fittings are quite expensive and making them up is tricky, time-consuming and tedious. The whole idea would be rather pointless, since standard CPVC plastic fittings are so much less expensive and easier to work with. This would be, however, an easy method of coupling together runs of PB tubing by using a flare-type coupling fitting. Also, the fact that these standard flare fittings can be used with CPVC and PB means that adapting either type to existing copper tubing plumbing lines is a simple matter (Fig. 4-15). In addition, either CPVC or PB can be simply connected to the many fixtures and appliances that are either equipped with or will readily accept standard brass flare fittings.

It's worthwhile knowing that these standard flare fittings are easily obtainable at all automotive and plumbing supply houses, as well as many hardware stores. On the other hand, sometimes plastic adapters or transition fittings are not so easy to come by. Standard brass flare fittings can themselves be obtained in a wide variety of adapters to fit various types and sizes of metallic piping, giving one a wide range of possibilities in cobbling up a fitting series to suit some particular problematical plumbing application.

COMPRESSION FITTINGS

Compression fittings have also long been used in making joints in copper tubing (Fig. 4-16). The fitting is made of machined brass, and consists of a fitting body along with an appropriate number of flange nuts. For each flange nut, there is an accompanying metal ring called a ferrule or compression ring. Though in relatively common use in automotive mechanics and other applications, they are less widely seen in household plumbing applications. Nonetheless, they are found with some frequency in certain stop valve applications, such as the chrome-plated ones often used in toilet, sink or washbasin supply lines (Fig. 4-17). The stop valve itself is designed to fit to a standard threaded pipe, while the riser tube leading from it may be joined with a compression fitting.

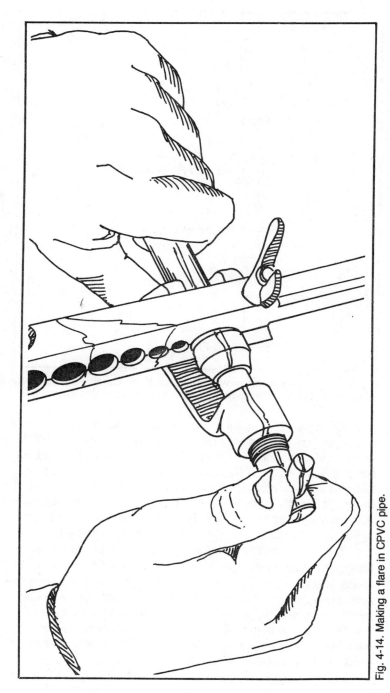

Fig. 4-14. Making a flare in CPVC pipe.

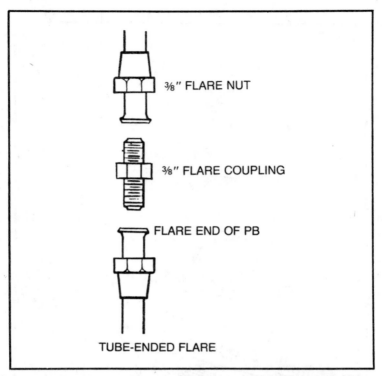

3/8" FLARE NUT

3/8" FLARE COUPLING

FLARE END OF PB

TUBE-ENDED FLARE

Fig. 4-15. Joining PB to copper with flare fittings. Numerous similar adaptations can be made by using various types of flare fittings (courtesy of Genova, Inc.).

Making a compression joint is not a difficult job, but must be done exactly right if it is to be leak-free. First, square the end of the tubing, and file it flat if necessary; both the inside and the outside edges should be free of burrs or ridges. Next, slip the flange nut, followed by the ferrule, onto the tubing and out of the way. Insert the end of the tubing into the fitting body as far as it will go, and slide the ferrule up over the tubing and into the fitting body. Follow with the flange nut, which should be taken down wrench-tight.

The danger with this type of fitting lies in over-tightening the flange nut, which throws the ferrule out of line and/or deforms the tubing or the ferrule. The ferrule cannot be removed, and the only recourse is to cut the end of the tubing off and start again with a new ferrule. In view of the facts that these joints can be troublesome and that there are so many other possibilities for making a good joint with ease, compression joints are just as well avoided wherever practical and possible.

NO-HUB FITTINGS

No-hub joints are so simple and effective in application that it is a shame that they are not allowable for use in many areas. The joint consists of a neoprene rubber sleeve covered by a stainless steel sleeve and held in place by stainless steel screw clamps of the automotive variety, just like those used to secure insertion fittings in polyethelene pipe. They are used primarily in no-pressure applications on DWV systems. They can be used with any kind of thick-walled pipe, either to join free sections of identical pipes or make the transition between one kind of existing pipe and another of the same size. In the former case, the neoprene sleeve has a shoulder centered around the inside, so that the fitting will only slide halfway onto each section of pipe. In the latter case, and especially when cutting a new branch line into an existing pipeline, slip sleeves

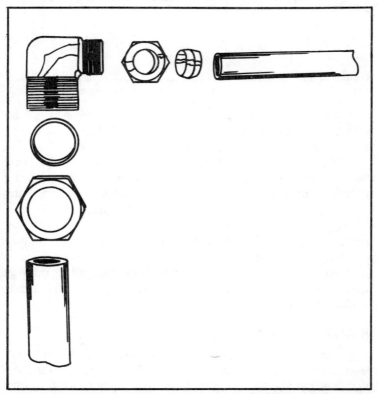

Fig. 4-16. Component parts of compression fitting. Right-angle adapter accepts ⅜-inch copper tubing with brass ferrule and compression nut at one end, and ½-inch CPVC pipe with plastic ferrule and brass compression nut at the other.

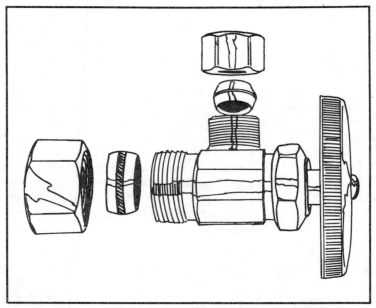

Fig. 4-17. Chrome-plated brass compression fitting angle stop valve.

that lack the centering shoulder are used. The sleeve is slid all the way onto one pipe until it is set in place (Fig. 4-18). Then it's slid halfway back to join with the other pipe (Fig. 4-19).

Making the joint is simple and practically foolproof. All one need do is slip the neoprene sleeve into position over the pipe being joined, wrap the metal sleeve around the rubber one, and attach the screw clamps and tighten them down with the maximum pressure. This type of joint has the further attribute of being flexible, so that pipe sections can be laid at slight angles to one another. The fittings are also simple to disassemble and are reusable. For the home plumber, of course, the chief advantage lies in being able to make an excellent joint without difficulty when connecting a new DWV piping section to an old one.

GASKETED FITTINGS

Gasketed fittings, of which there is a considerable variety, are used here and there for particular applications throughout the average residential plumbing system. The gaskets take various forms, sizes and shapes for different purposes within different kinds of fittings. However, they all work much the same in principle, by expanding slightly to form a tight and leak-free seal as they are

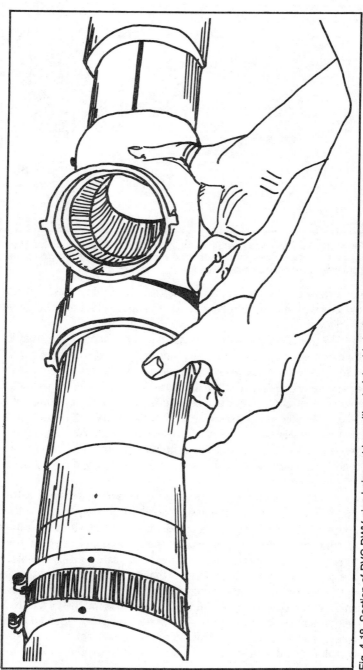

Fig. 4-18. Section of PVC-DWV pipe being set in position to join old pipeline with No-Hub sleeve type coupling.

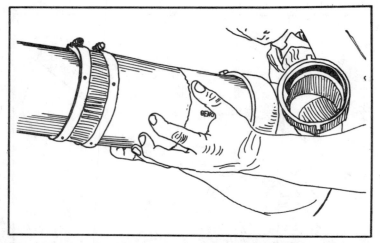

Fig. 4-19. With old and new pipes properly aligned, No-Hub sleeve is slid forward to cover the joint between the pipe sections and the clamps are tightened to complete the installation.

compressed within the fitting and at the joint. Metal rings, washers made of various substances, sleeve rings and O-rings are commonly used.

In the DWV system a special type of gasketed joint may be found on existing cast-iron lines, and on PVC sewer lines as well (Fig. 4-20). They are also widely used on both metallic and plastic drainpipe fittings such as the slip nuts and adjustable trap arrangements mentioned earlier in the discussion of threaded fittings (Fig. 4-21). Metallic waste pipes beneath washbasins and kitchen sinks also use gasketed fittings, and a washer-type gasket is used between the metal drain and the sink body itself. In addition, many types of adapters and transition fittings include gasket seals of one sort or another.

The installation of fittings of this type is a very simple matter, and for the most part there are likely to be only two points of concern. One is that the gasket itself must be properly centered and aligned within the fitting so that it seats correctly, can seal effectively, and does not become distorted or damaged by misplacement. The second consideration is that the job of tightening should not be overdone. Most gaskets compress and seal with a relatively small amount of pressure. All that is needed is for the gasket to be seated sufficiently tight enough that no leakage or seepage takes place. This is a more critical situation, of course, with pipelines carrying liquid under pressure than it is with drain lines. In a

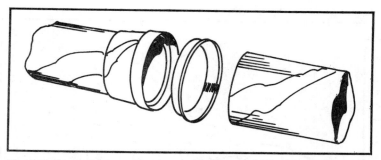

Fig. 4-20. Gasketed sewer pipes are slipped together and the gasket ring seals the joint.

pressurized line, gasketed fittings should be taken down firmly snug, but never forced. Gasketed fittings in drainage lines usually need be no more than hand-tight plus ¼-turn. If that proves to permit an occasional bit of seepage, another ⅛-turn will probably make the seal.

WELDED JOINTS

Most of the joints that you make while assembling a plastic plumbing system, both in the water distribution and the DWV

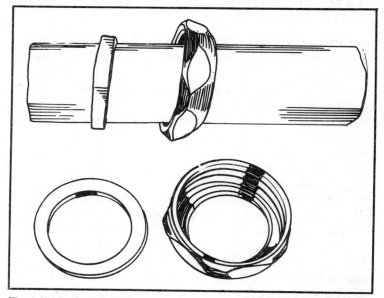

Fig. 4-21. Gasketed slip nut joints are commonly used in trap assemblies and tubular product components.

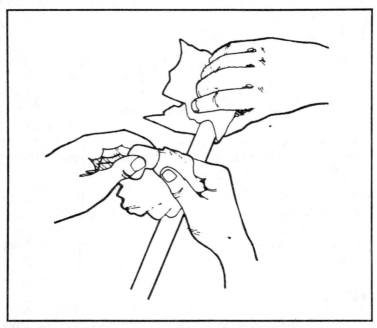

Fig. 4-22. First step in solvent welding, after cutting pipe and dressing ends, is to thoroughly clean pipe and fitting socket (courtesy of Genova, Inc.).

systems, are likely to be of the solvent-welded type. This is perhaps the easiest and one of the most effective of all piping joints to make, and therein lies a potential problem. The whole process is so simple and so fast that there is a tendency to become overconfident and just slop things together without paying proper attention to the process at each individual joint. The inevitable result may be seepage, leakage of joint failure, and extra work and aggravation for the plumber. These problems can be neatly sidestepped by making each joint carefully and in the proper manner, using the right supplies and techniques.

The process of solvent-welding is commonly used with PVC, CPVC, ABS and SR pipe. No matter what the type, the pipe ends must first be clean. They should be wiped free of dirt, mud or moisture, and then swabbed thoroughly with a proper cleaning solution (Fig. 4-22). This fluid reacts slightly with the plastic and serves not only to clean off residual dirt, but also grease, oil and fingerprints that might interfere with the welding process. You can use the type specifically recommended, and perhaps supplied by, the parts manufacturer, or you can use one of the all-purpose cleaners that will do a satisfactory job on the three types of plastic pipe. Apply the cleaner from a saturated paper towel or rag, and allow the pipe end to dry thoroughly. This only takes a moment. Make sure that you have cleaned well beyond the point where the end of the fitting will lie when the joint is made up. After cleaning, don't touch the pipe end with your fingers, or lay the pipe in the dirt. Clean the inside walls of the fitting sockets in the same manner.

The next step is to apply the welding cement. Again, you have a choice of materials that can be used. Some cements are meant to be used only with CPVC pipe and fittings, others only with PVC, and some are designed for ABS. None of them should be interchanged. However, there is also an all-purpose type of cement that is suitable for PVC, CPVC, ABS and SR pipe types. In addition, you can obtain a special cement for use in low temperatures. This cement, however, is not used with CPVC; it does work with the other three types.

The Welding Process

The process of welding is simple enough. Leaky joints are almost invariably caused by skimping on cement, so don't be afraid to give each joint an ample dose. Most cements are sold in screw-top cans with a brush or dauber attached to the inside of the cap,

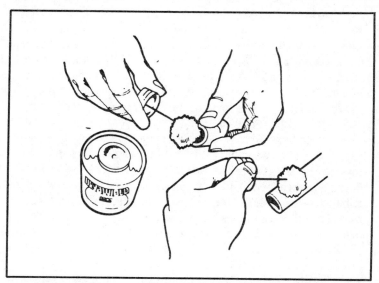

Fig. 4-23. Solvent cement is applied heavily to pipe end, and in thin coat in fitting socket (courtesy of Genova, Inc.).

which simplifies matters. Smear a liberal amount of cement on the pipe end, making sure that you get past the point where the end of the fitting will lie. Immediately coat the inside walls of the fitting socket, but not too heavily (Fig. 4-23). Stick the dauber back in the can and slip the two pipes together with no wasted time (Fig. 4-24). As you push the pipe into the fitting (or vice-versa) give the fitting a slight twist, no more than a quarter-turn. Be sure that the pipe is all the way home in the fitting. After the slight twisting motion and the seating of the fitting, do not move the two parts again; hold them firmly aligned for 5 seconds or so, a bit longer if they are under a slight pressure or strain that might cause them to misalign.

If the fitting must be set at a particular angle, as is often the case, don't rely upon the judgement of your eye to get the correct positioning. Before you even clean the pipe and fittings, put them together dry, set the fitting at the correct angle, and make a pencil or scratch mark on both pipe and fittings. Then when you weld them together you can quickly and easily align the two marks during the process of twisting the fittings onto the pipe. You get a perfect lie every time.

Points To Remember

There are a few other points of interest about this process, too. For instance, if you build up a small ridge of wet cement as you push the fitting into place on the pipe, that's good. There should be an obvious bead all the way around the end of the fitting. If there seems to be an excessive amount, perhaps dripping off the pipe, don't worry about it. No harm is done (but don't let the stuff drip on any finished surface). An excessive buildup of solvent cement at the base of the fitting socket inside does tend to cut efficient water flow to some degree; make this coating of cement relatively thin.

Another point is to do your work in warm weather. Atmospheric conditions do have an effect upon the solvent-welding process. For your own comfort and for the sake of the joints you are making up, try to work in ambient temperatures of at least 50 degrees Fahrenheit; CPVC pipelines for pressurized water distribution service should be made up in temperatures no lower than that. The other pipe types can be put together successfully with the aid of the special low-temperature welding solvent cement mentioned earlier. Even so, the job is probably better done during warmer weather. High temperatures also have an effect upon the rapidity with which a joint cures, and so does humidity. In short, the idea of conditions are the same as those we all prefer to live and work in—about 70 degrees Fahrenheit and 45 percent or so humidity. Cold, however, is the important element to watch out for.

A third point, minor but important, is to make sure that no dirt, grit, sawdust or other foreign matter blows or falls onto the wet pipe joint as you are welding the parts together. Anything that touches the cement will stick fast and get crammed into the joints. This could possibly have a detrimental effect on the joint seal.

Last but not least, you will find that solvent-welded joints set up rather quickly. This means that you don't have time to fiddle around with them. You must get on with the job of joining the parts as quickly and efficiently as you can, but not hastily enough to louse something up. You do have a minute or two of usable time. But just because a joint sets up rapidly does not mean that it is immediately ready to go into service. It's a good idea to allow pressurized pipeline to cure for at least an hour for low pressure, longer for higher pressures, before putting them into service. A longer period of time is even safer, because by the time the joint is a day old the pipe will rupture before the joint will separate. Also, if you are solvent-welding a large-diameter drainpipe section together on the ground surface preparatory to lowering the whole affair into a trench, let the completed pipeline lie still overnight, or for at least 12 hours, before moving it. This will allow the joints to cure sufficiently that no damage will be caused by the considerable stresses and strains imposed upon them when moving the line around.

FIXING MISTAKES

Making a mistake or two during the course of completing a residential plumbing system is almost inevitable. It's awfully easy to

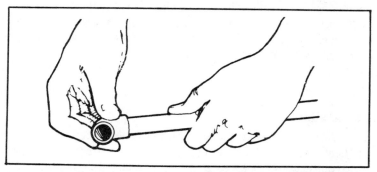

Fig. 4-24. As soon as solvent cement is applied, slip pipe and fitting together in a slight turning motion until the fitting seats, then hold for a few seconds (courtesy of Genova, Inc.).

do. An ell fitting gets turned the wrong way, a pipeline gets routed on one side of a framing member instead of the other, or maybe the original calculation for a piping section was incorrectly done in the first place. With metallic plumbing systems, the offending section is disassembled with relative ease by unscrewing fittings, unsoldering or whatever. When plastic pipelines are solvent-welded, though, the joint is a permanent one. It can not be taken apart. What then?

This is not as much of a problem as it might seem at first glance. Be aware that indeed there is no way that a once used solvent-welded plastic fitting can be used again. Even if you discover a mistake just as you are joining pipe and fitting and can pull the fitting off again, there is no point in trying to use it again elsewhere. Don't take a chance; throw the fitting away. But there is an easy way to correct mistakes. Just cut the offending section of pipeline right out of the system at whatever seems to be the likeliest point. Since plastic pipe works so easily, this only takes a moment and not much effort. Trim any usable sections of pipe from the welded fitting, and throw the fitting away. Save the intact section for installation in another part of the system yet to be assembled. With the aid of a couple of extra coupling fittings, insert a replacement section correctly put together (Fig. 4-25). The section you removed can perhaps be grafted in elsewhere later on, again with the aid of a coupling or two or other fittings.

If the problem should be an ell cocked in the wrong direction, cut the pipe off a few inches in back of the fitting, splice in a coupling and twist the pipe in to the appropriate direction as you are putting the new joint together. In short, by cutting the pipeline and inserting new bits and pieces as necessary, you can actually make repairs, additions or corrections in a plastic pipeline with a minimum of time, difficulty and expense. With some careful head scratching, many if not all of the removed piping sections, parts and/or fittings can be coupled back into the system at other points. Don't throw anything (except a half-welded fitting) away; something might come in handy later.

EXPANSION PROBLEMS

All substances expand with the application of heat and contract as heat is extracted. Plumbing systems are no different. Expansion and contraction of pipelines in metallic residential plumbing systems is seldom given much consideration, because the degree of movement of the pipelines is usually negligible from a practical standpoint. With plastic piping, however, the situation is somewhat

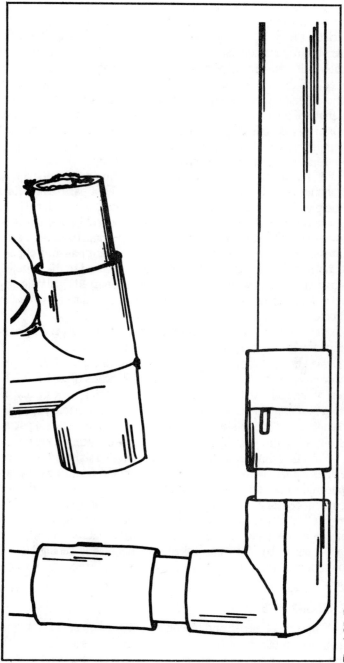

Fig. 4-25. Repairs or changes can be easily made by sawing out the unwanted section and welding in new fittings and pipe sections.

179

different. As mentioned earlier, plastic will expand in diameter, especially the flexible type, when water freezes within the line. Expansion in diameter also takes place in hot water distribution lines as cool, still water lying in the lines quickly heats up when a tap is opened and the flow of hot water begins from the water heater. In addition, plastic pipelines will expand a considerable amount over the length of the pipe. All plastic lines will expand both in length and diameter to a certain degree with the normal fluctuations of outside air temperature and inside water temperature, or with the varying and often quick temperature changes that occur in drainpipes.

For the most part, not much consideration need be given to this phenomena. However, there are a couple of particular points for the plumber to keep in mind. One was mentioned earlier. Holes bored in framing members to carry either hot or cold water distribution lines should be bored oversize, well-aligned and trimmed free of splinters. Otherwise, the lines will creak and chatter as they expand and contract. By the same token, pipelines secured to the edges of joists or other framing members should be supported by plastic pipe straps that fit somewhat loosely and allow the pipeline to slip easily and quietly.

Another point is that runs longer than 10 feet of rigid plastic piping, whether hot or cold water supply, should include doglegs. A dogleg is simply a jog made in the pipeline with a pair of 90-degree ells, creating an offset in the line of about 12 inches (Fig. 4-26). The inherent flexibility of the material allows the dogleg to flex back and forth as the pipeline changes length because of thermal expansion and contraction. It constitutes a sort of relief valve, releasing strain and obviating the possibility of any pipeline or fitting damage.

Note that this need only be done in long straight lines. Where the lines contain numerous directional changes and fittings, and are supported freely and not jammed into corners or whatever, doglegs are not necessary. The pipeline segments set at various angles to one another fill the same purpose. The point, then, is to make sure that the entire pipeline system has plenty of latitude in which to move around. By the same token, flexible plastic tubing like PB should never be pulled taut. Allow it plenty of freedom of movement.

MEASURING PIPE

The correct measuring of lengths of rigid pipe seems for some reason to contain a lot of pitfalls for many home plumbers. The principal reason for this, one suspects, is a simple matter of not

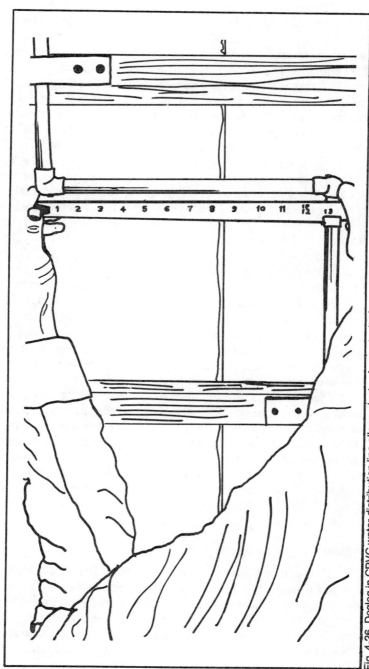

Fig. 4-26. Dogleg in CPVC water distribution line allows plenty of opportunity for expansion and contraction of pipeline without damage or noise.

181

paying attention to the job at hand. Measurements must be taken with care, especially where quarters are cramped and distances between fittings are critical. Pipe sections must always be fully seated within the fitting socket; otherwise the joint will be less than fully effective. Miscalculations in measurements can also lead to wasted fittings and pipe pieces, unnecessary strain on the pipeline if miscalculated sections are made up anyway, and extra time, trouble and frustration for the plumber, not to mention a sloppy-looking job.

The crucial point to remember is that the length of a section of pipe is equal to the distance between the ends of a pair of fittings, *plus* the amount of pipe that must slide inside the fitting socket at each end. One must also take into consideration such factors as the width of a shoulder in the middle of a coupling, or the distance that is offset by the directional turn of a particular fitting. Another important factor is to make sure that measurements are made from accurate base points; this should not be done by eye, but with the aid of level and plumb bob. Many of your measurements will be made "in mid-air," so to speak. Without establishing accurate reference points, there is no way to obtain accurate measurements.

For a home plumber who is unfamiliar with the process of measuring up pipeline sections and making the necessary allowances for fittings and such, one of the easiest methods to follow is to lay all of the parts for a particular pipeline segment out on the floor. Then make transfer measurements as necessary, marking pipe sections for cutting as indicated.

There really is no way to explain exactly how to go about making the required measurements, because the process is dependent to a great degree not only upon the specific pipes and manufacturers of the pipe and fittings being used, but also upon the specific circumstances of installation. About all that can be said is that this is an area where problems can quickly arise through improper or incorrect calculations. The best bet is to proceed cautiously and double check all calculations and measurements by whatever means seem most logical, given the particular conditions.

COMPATIBILITY OF PARTS

The statement has been made elsewhere in this book that all of the piping and parts in a given plumbing system or subsystem must be compatible. This is such an important factor in the sucess of a plumbing system that it bears repeating. There are numerous manufacturers of plastic plumbing products. In a metropolitan area where there are likely to be several well-stocked plumbing supply warehouses, the plumber has a grand array of choices in pipe, fittings and various accessories. Many of the parts made by one manufacturer can be interchanged with those made by another, at least insofar as gross sizing, apparently satisfactory fit, general appearance and similar factors are concerned. Indeed, frequently parts and pipes of different brand names can be successfully interchanged. The problem is that you never know which parts can be successfully interchanged and which cannot. There are plenty of the latter that are apparently all right when the system is made up, but not potentially capable of withstanding long periods of service.

Several reasons exist for this, but one of the principal ones is the wide range of size tolerances allowed in the manufacturing standards. There may be considerable variation in the out-of-roundness of pipes, a range of wall thicknesses and corresponding inside or outside diameters, and a similar span of sizes in fittings made by various manufacturers. Thus, if you happen to fit a skinny pipe to a fat fitting, a weak joint could result. Incidentally, these differences in tolerances are measured in thousandths of an inch, and are not discernible to the naked eye.

Note, however, that the problem is compounded by the fact that some manufacturers make only piping, while others make only fittings. This means that you may not have a choice in buying, but must use whatever is locally available. If this is the case, use the combinations of pipe manufacturers and fitting manufacturers that have proven successful in your area, as determined by local practice and experience. The best bet, of course, is to buy pipe and fittings produced by the same company.

Another problem in compatibility lies in the different types of plastic materials. Fittings for use with one type of pipe should not be interchanged with those designed for another. Even if they might fit, the joint is unlikely to be a good one. Always use compatible pipes, fittings and accessories that are designed to complement one another.

DRAINAGE SLOPE

Arranging the correct drainage slope in a pipeline, whether for household drainage purposes, groundwater drainage, leaching field or simply so that a water distribution system can be effectively drained during cold weather, would seem to be a perfectly straightforward task. However, such is often not the case. More will be said later on about the particular correct pitches that various kinds of pipelines should follow. Determining a proper pitch seems not to be a difficulty, but installing the pipeline at the proper angle and maintaining the pitch in a continuum throughout the pipeline does.

Whatever particular pitch a pipeline follows, that pitch should remain constant (except in the case of sharp-angle drops in waste lines and the like), and the pipeline should never be allowed to belly or sag downward at various points along the line. Maintaining the proper pitch from one end of a pipeline to the other is a matter of making the correct initial calculations. If a line is 40 feet long, for instance, and the pitch runs downward at the rate of ¼-inch per running foot of pipeline, obviously one end of the line must be 40/4-inches, or 10 inches, lower than the other. Setting each end of the pipeline at the proper relative point should pose no problem, but somehow often does.

Nor should the pipeline be allowed to sag at any point along the way. The bottom of the pipeline should lie as straight as a stretched string. This is a matter of providing adequate support. A series of closely spaced pipe hangers, pipe straps, brackets or other supports spaced at short intervals along a pipeline will do the job. Where applicable, a grade board staked into solid subsoil does an even better job. When necessary, the grade board principle can be employed by nailing a properly graded runner board in place within a building to provide continuous support, and then securing the pipeline to the runner board.

Where an entire water distribution system must be installed in a residence so that the system can be completely drained out during cold weather, the lie of each individual segment of the system must be carefully lined up and installed so that all slope back in a continuum to a central point where the drain valve is located. Sometimes this is a difficult proposition, and the choice is made to arrange the lines so that they can be blown out with compressed air instead of attempting to install an involved series of correct pipeline pitches.

ROOF JACKS

A roof flashing unit, widely called a roof jack, is a special device designed to slip over a plumbing stack where it protrudes through a roof, in order to prevent leakage of rainwater into the building or under the roof weather surface. Roof jacks would hardly need mention, except for the fact that so many of them are improperly installed. Though this has no effect upon the proper operation of a plumbing system, a leaky flashing job certainly can cause substantial aggravation, expense and extra work for the homeowner.

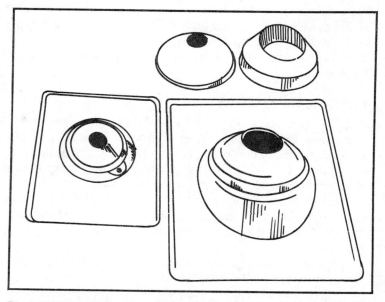

Fig. 4-27. Pipe flashings or roof jacks provide a weathertight seal around stacks that protrude through the roof.

Roof jacks are commonly made of galvanized sheet metal, but the best kind for use with a plastic plumbing system is a special thermoplastic type made for just this purpose (Fig. 4-27). The plastic is completely weatherproof, and installation of a jack is a simple chore. The trick is to do it right the first time.

On a new roof, the hole for the vent pipe can be cut as soon as the roofing underlayment is laid upon the sheathing, and before the roof weather surface is applied. The hole need only be large enough to admit passage of the pipe, which can then be set in place. Then the thermoplastic sleeve is pulled down over the pipe. The material is elastic, and will stretch snugly around the pipe wall. The jack is pushed down until the base rests flat upon the underlayment. Some plumbers apply a fairly thick layer of roofing cement or other sealant to the upper two-thirds of the underside of the flange, as viewed looking up the roof pitch towards the ridge line.

When the roof weather surface is applied, the material is slipped beneath the lower one-third of the jack flange, but over the upper two-thirds. This allows rainwater to flow down the roof and over the jack flange without creeping underneath. The exposed edge of the flange, the lower part, can also be sealed down to the weather surface if desired. In no case, however, should the entire jack flange be located beneath the roof weather surface, whether sealed down or not. By the same token the entire flange should never be located on top of the roof weather surface. Neither of these two methods will provide an adequate weatherproof seal for the vent stack.

SECTIONAL DRY ASSEMBLY

One of the best ways to avoid any immediate difficulties or later problems in the assembly or later operation of a plastic plumbing system is to employ the process of sectional dry assembly as the job goes along. This is particularly advantageous for the beginning plumber, since consequential mistakes can be eliminated with only a minor loss of time. Though the process initially adds a certain amount of work time to the project, the overall result is likely to be more satisfactory and quite possibly in the long run a time-saving procedure.

Instead of starting at point A with a piece of pipe, welding on a fitting, securing the pipe, welding in the next section and securing it, then going on to the next fitting and section of pipe and so on, completing the system bit by bit, use a different approach. Establish the first piece of pipe at a convenient starting point in the system, or section of the system, and put a fitting in place dry, without welding. Make sure that the fitting is firmly seated and cut the next piece of length, insert it into the fitting and secure the pipe. Put the next fitting on and go on to the next piece of pipe. At no point is the system actually permanently installed. This is just a dry run.

Once an entire segment of logical proportion is put together and checked out for proper fit, lie and so on, go back to the beginning and start over again. Disassemble the dry pipe and fittings, weld them together, piece by piece, and secure the pipes permanently, in the same order as you originally put them together.

The one thing that you must be sure of during the dry-run assembly is that all pipes and fittings are completely seated. In quality piping products and especially with CVCP water supply components you will probably encounter what is called an interference dry-fit, which can lead to problems in dry-run assembly. An interference dry-fit occurs when the pipe wall contacts the fitting socket wall from one-third to two-thirds of the distance into the socket. This makes for the best joint, but if full seating does not occur during the dry-run assembly, dimensional discrepancies in pipe runs can result. Check this carefully; if full seating is impossible, make the necessary allowances in pipe section or overall pipe run lengths. Otherwise, when you apply the welding solvent, it can act as a lubricant that allows you to fully seat joints that possibly were only partly seated when dry, leaving you short a bit of pipe here and there. This in turn might mean that the segment will not properly fit in its appointed location.

But all in all, even though the dry-run process if incorrectly done can lead to some assembly problems, this method is a good one for the later avoidance of more major problems that could be more difficult to correct.

FIXTURE CONNECTIONS

Final connection of a pipeline, whether water supply or drainage, is usually one of the last steps in making up the plumbing system. Often the pipelines are merely stubbed out of the wall or up through the floor, with the final connection being made only after the rest of the pipeline has been installed (Fig. 4-28).

Fixture connections are made in a great many different ways, depending upon the specific fixture involved. Drainage connections, except for toilets, are made by attaching a trap of one sort or another directly to the fixture drain outlet and to the drainpipe line by means of a suitable tail pipe and a slip nut. The toilet connection is made at the same time the toilet is installed, by positioning a wax ring especially made for this purpose atop the closet flange, setting

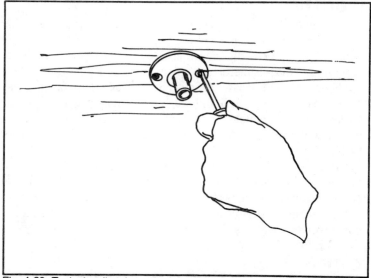

Fig. 4-28. Typical wall stub-out for later fixture attachment.

the toilet atop the ring and bolting the fixture down. The toilet water supply connection is generally made with a ⅜-inch riser tube assembly which connects directly to the toilet tank inlet at one end and to the water supply pipeline or perferably a stop valve in that line. Compression fittings are often used here, but there are other arrangements as well.

Faucet assemblies, whether for washbasins, kitchen sinks or laundry tubs, are generally made in the same fashion, with riser tubes running from stop valves to the faucet base. Hose bibs or sillcocks can be adapted directly to the plastic water supply pipeline, and in fact solvent-welded plastic units can be directly attached. All of these connections are made by various means and with different kinds of piping and adapter or transition fittings. Any fixture can be connected to any kind of pipeline under virtually all physical circumstances simply by using the correct fitting.

The principal point to keep in mind when making these connections is that for good service they must be carefully and correctly installed, tested and found to be leak-free before they are covered up or hidden by finish material. Accessibility and the will to get the job done right is always much greater during the time the job is underway than it is after the whole project has been finished up and the tools put away.

CUTTING PIPE

Cutting plastic pipe is a simple process, much easier than working with metallic pipe, but for clean joints and a good fit the job must be done correctly. There are several methods that can be employed. Small-diameter rigid pipes such as PVC or CPVC can readily be cut with a saw. A fine-toothed blade is by far the best, since the coarser the blade the rougher and more ragged the cut. A hacksaw with a metal-cutting blade works fine, and so does a backsaw or miter saw (Fig. 4-29). Remember, though, that most types of plastic pipe have a dulling effect on ordinary sawblades designed primarily for cutting wood and other soft or nonabrasive materials. If a great many cuts must be made, consider a metal-cutting blade in a power jigsaw or circular saw. A table saw or radial arm saw can also be used, provided there is sufficient space to maneuver the sections of pipe about.

It is important that the pipe be squarely cut, and not canted off at an angle. This happens automatically with power saws, but when cutting by hand, set the pipe in a miter box so that you can get a perfectly square cut. Also, make sure that the pipe is firmly and

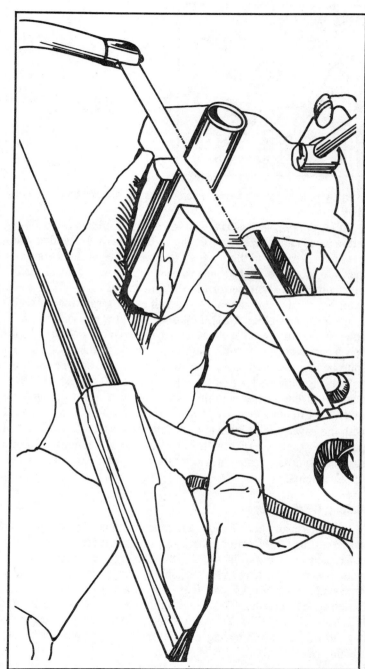

Fig. 4-29. Rigid plastic pipe can be cut easily with a hacksaw fitted with a fine-toothed blade, or with any other fine saw.

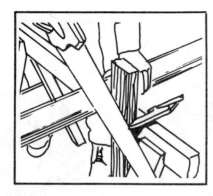

Fig. 4-30. Large DWV or sewer pipe can be supported on sawhorses and wedged against a block, then cut with carpenter's crosscut saw.

securely held in place so that it can not move around while you are sawing.

Larger pipe, such as DWV or sewer pipe, can also be easily cut with a saw. Large sizes are too big to be effectively handled with most power saws, so the best bet is a handsaw. A deep-throated hacksaw works well, but an ordinary carpenter's handsaw in a fine-toothed crosscut pattern is even better (Fig. 4-30). A deep miter box, which you can make yourself out of scrap wood, greatly helps the process (Fig. 4-31). Make sure that the pipe is firmly held down and can not wobble about, and make each cut slowly and with only a minimum of pressure. Let the saw blade do the work. As you approach the end of the cut, slow the cutting action down even more so that the pipe wall will not crack or chip away as the last bit of material is cut. When working with thin-walled sewer pipe, be careful not to put too much pressure on the end of the pipe, either with the saw blade or by holding it down too firmly. Some kinds of pipe can crack or split with surprising ease, especially if the sawblade binds and catches in a certain way, putting too much strain upon the pipe.

Cutting CPVC Piping

There is another method, an easier one, that can be employed when cutting CPVC water supply piping. A tubing cutter will do the job rapidly and easily (Fig. 4-32). The standard type of cutter that is used for copper tubing and is available at any hardware or automotive supply store works well. However, it will also leave a distinct, rolled-up ring all around the outside of the cut, as well as a lip around the inside (Fig. 4-33). To avoid this problem, use a specially designed tubing cutter having a very thin blade capable of cleanly cutting plastic pipe.

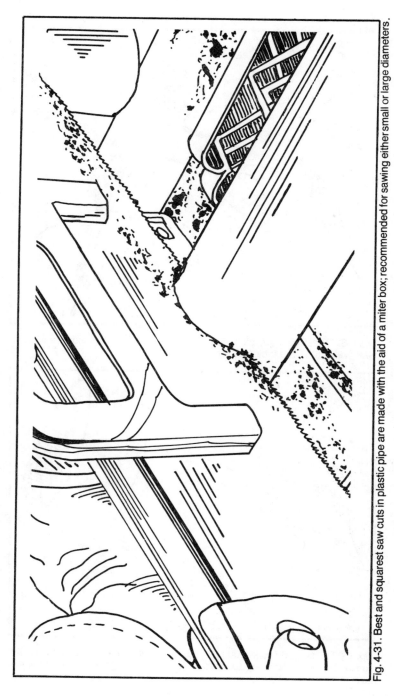

Fig. 4-31. Best and squarest saw cuts in plastic pipe are made with the aid of a miter box; recommended for sawing either small or large diameters.

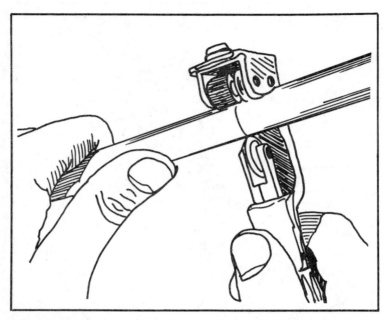

Fig. 4-32. Small diameter pipe, such as this ½-inch CPVC, is most easily and effectively cut with a tubing cutter.

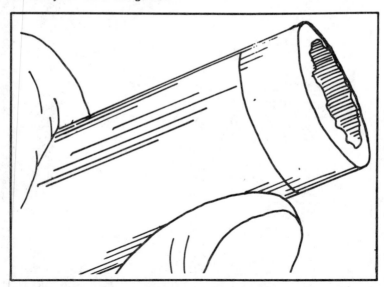

Fig. 4-33. Standard copper tubing cutter or pipe cutter leaves rolled ridge on outside of pipe end, and a ragged rim of plastic on inside. Both must be trimmed away.

Once the pipe has been cut, the end must be dressed. If saw cutting has left a ragged face on the pipe wall, file it flat with an ordinary medium-toothed metal file. An outside ridge left by a copper tubing cutter can likewise be filed off by holding the file flat against the teeth of the file (Fig. 4-34). File the ridge off flush with the pipe wall, but be careful not to chamfer or bevel the end of the pipe. Burrs or a ride left around the inside of the pipe end by cutting can be trimmed away with a pipe reamer, a rattail file or a sharp knife. If you use a knife, though, do so gently so that you do not carve chunks out of the pipe. However you do the job, close attention should be paid to making the pipe end perfectly trim and clean, in order to effect a good joint.

Cutting Flexible Material

Flexible plastic pipe that uses insertion fittings, such as polyethelene, can be cut with either a hand saw, a power saw or a

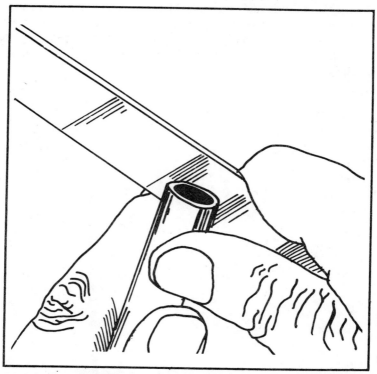

Fig. 4-34. Rolled-up ridge on pipe cut with tubing cutter can be filed away by rolling the edge of the cut against a file.

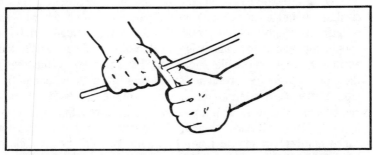

Fig. 4-35. Relatively soft and flexible plastic pipe or tubing like PE or PB can be easily cut with a razor-sharp knife.

very sharp knife. The outside edge of the cut end is of little concern, and cutting the end exactly square is not as critical in this situation as it is with other types of pipe. However, the inside edge of the cut end should be trimmed clean of burrs, so that the insertion fitting will seat properly and no chunks of plastic tubing, like PB, can be cleanly cut with a very sharp knife (Fig. 4-35). With the larger sizes, a very fine-toothed hacksaw might also be used. Again, the end should be clean and trim to allow the "grab fittings" to be easily installed.

Read through them thoroughly to find out what portions or particular laws or ordinances are likely to affect your plan program.

In particular, look for references to building codes of whatever sort. In some cases entire national building codes are adopted *en toto*. In others, only certain sections of national building codes may be adopted and listed by reference numbers. In yet other cases, a separate body of local building code provisions may have been drafted especially for use in that area. If short, they may be included in the zoning laws. If not, they will found elsewhere. Whatever the case, if references to other bodies of regulations are made, you must then track those down and peruse them to see what effect they might have on your program.

Know Officials' Attitude

Once you have the information in hand, the next step is to discover just how much use those in local authority make of the rules and regulations. Often they are strictly and stringently enforced with a bevy of inspectors and other officials. There will be teeth, sharp ones, behind the laws. In other cases, you may find that the whole program is rather lax, perhaps even to the point of being laughable, with little or no enforcement taking place. The best way to find our just which way the wind blows is to talk with you neighbors, friends and especially local contractors to find out just what actually goes on in the area in the way of following zoning provisions, and just what can be done and what cannot.

This is not a matter of attempting to "get away with something," or to beat the system. Such shenanigans generally are self-defeating anyway sooner or later. Rather, it is a case of determining exactly what must be done and what latitude you have within which to work. You can thus do a good, effective job (in this case of plumbing installation) at optimum cost in cash, time and effort, and at the same time stay within the bounds of effective and purposeful local regulations and custom. Local zoning ordinances may specify one thing. If half of the area inhabitants have opted for another in apparent violation but officialdom has turned a blind eye to the situation, or relented in the face of citizen opposition without bothering to change the pertinent law, there is no reason why you should be the lone soul to be in compliance.

Check carefully. There is another situation that sometimes develops, a condition known as "grandfathering". All this means is that the folks no longer in compliance with a particular regulation, because that regulation was passed after the fact of their current noncompliance, are exempt from the regulation. Their situation was perfectly   legal at one point and became seemingly illegal after passage of law. Thus, while all the long-time residents have no concern, a newcomer following the same pattern or procedure of building, or whatever, might willy-nilly find himself in difficulties with the authorities for noncompliance.

CODES

The actual process of construction, be that industrial, commercial or residential, is often governed by building codes of one sort or another. There are three in particular that you will hear about: the Uniform Building Code, the National Electrical Code and the National Plumbing Code. These codes have been developed, and are being constantly revised, with an eye toward national standards in the construction industry. Some locales have adopted these codes just as they stand. Others have excerpted certain portions of the codes that they happen to find satisfactory and applicable for certain purposes in those areas. Some communities have chosen to use bits and pieces of the code and to write up additional articles as necessary to cover local needs. The result is a mishmash of rules and regulations to be found all over the country, frequently with no particular coordination or cohesion. Of course, there are still many areas where there are actually no codes in effect whatsoever, though some argue that where this is the case, building construction should be guided by the various national codes anyway.

Plumbing Code Compliance

Our primary interest here is the plumbing code, whether national or local and of whatever variation, content or length, as it affects the individual homeowner plumber. What is said of the plumbing codes can likewise be said of the other codes. There are many who rail against plumbing codes for one reason another, and they undoubtedly have some good points. The plumbing codes have gained their greatest impetus from the various factions that have the most to gain by establishing and strictly enforcing such a body of laws. These factions include tradesmen who profit by them, manufacturers who prefer to see their materials used and not some other kind, governing bodies that collect fees and keep tight control, artisan groups and associations who prefer to keep their methods and techniques under wraps and lessen any potential competition, politicans and bureaucrats who gain in both jobs and funds through the strict regulation of building, and so on.

There is little question but what complete compliance with involved plumbing codes does indeed elevate the cost of a given job over what it need be. The codes do hamper and restrict the individual home plumber in his work and create many problems regarding what materials, procedures and techniques can be used, more than is really necessary.

Because of rapid developmental work in plumbing systems materials, procedures and techniques, most plumbing codes are outdated as fast as the articles are written. Frequently it takes them years to catch up. This creates even more problems, because so many plumbing codes leave far more unsaid than they spell out. This in turn leads to judgmental decision on the part of individuals concerned with enforcing the code, folks who often as not are actually ill-equipped to make those decisions, through little fault of their own. And that, of course, leads to even more difficulties for the lowly homeowner who is merely trying to get permission to run a line of PB from a handy water supply pipe to a new lavatory.

There is another side to the coin, too. In the first place, large portions of the plumbing code have to do with various commercial and industrial installations that have no bearing upon the single-family residence. Other sections are concerned with the quality of and the standard by which a vast array of plumbing equipment and materials is manufactured. As far as the home plumber is concerned, this is unlikely to be of any concern since the materials that he buys at a reputable plumbing supply house will more than likely be made to approved standards, and are likely also to be approved for use in his community.

For the most part, the remaining portions of the plumbing code that apply to residential application have been written in for some particular reason and with some particular purpose in mind. The primary underlying concern is on of both public and individual health and safety, particularly with regard to the disposal of wastes. Much of the material pertinent to residential plumbing systems has to do with the correct installation of DWV systems, sewer line hookups, septic tank installations and such.

Although no plumbing code is meant to be a training manual for neophyte plumbers, reading a code can be an education in itself. This is so because the articles are a distillation of plumbing knowledge that has evolved over the years through practical application, experimentation, field research and the like. Materials, methods, techniques, variations and alternatives are all here. If you are able to read between the lines, so to speak, you will learn a great deal about how plumbing systems are put together in an effective (and legal) manner.

How Plumbing Codes Affect You

How does the plumbing code affect you? Even if you live in an area where there is no code whatsoever in force, a plumbing code, primarily the National Plumbing Code, does have an effect upon the way you put your system together. This is because many of the precepts in the code are not just law in certain areas, but have been found over the years to be the best way to go about some particular installation or segment thereof. You will find some of them in this book, and you will find them in any book on plumbing, whether explicitly expressed as such or not. For example, every plumbing fixture must have a trap. That statement, in effect, is part of any plumbing code. And code or no code, anybody who installs fixtures without traps is just asking for trouble.

Consider another example. The minimum allowable size for a building drain, horizontal branch, vertical soil or waste stack that serves a toilet is 3-inch diameter DWV pipe. And why 3-inch? Smaller diameters have been found to not work well in numerous instances—3-inch and larger does. For the most part, by following accepted practices and using standard materials in a proper fashion, a home plumbing installation will largely follow the National Plumbing Code anyway.

Plastic plumbing materials are widely approved all over the country by various code writing bodies and substantial numbers of governing authorities. Various types of plastic pipe and fittings have been used for numerous purposes: water supply, water distribution, sewer lines, DWV systems, ground water drainage and a host of commercial-industrial purposes for several decades. Under proper application and installation, the track record for these products has been excellent. Those materials designed for residential installation (and others as well) conform to the various proper standards, such as those of the American Society for Testing and Materials, and are approved by assorted government agencies, such as the Federal Housing Administration. There is nothing strange, mysterious or unusual about them; they are legitimate products that work effectively in legitimate applications.

Obtaining a Variance

You may find yourself in the situation where your local plumbing code disallows the use of plastic plumbing components or allows restricted use for certain things such as ground water drainage, septic tank leaching fields, sewer lines or whatever. If there simply happens to be nothing in the code that specifically permits the use of plastic materials and thereby gives the inspector a perfect opportunity to say no, by all means investigate the local process of obtaining a variance to the code.

Obtaining a variance to the local building code provisions is not usually a terribly difficult chore, though it will require some thought, energy and time. There will be certain specific procedures which must be followed. You can obtain the details from the building inspector or from a member of your local Planning and Zoning Commision or Board of Appeals. You may choose to seek a particular variance that is applicable only in your own case—permission to install a CPVC hot and cold water distribution system, for instance. You may choose to attempt to have the existing building code rewritten in certain parts or added to, provisions specifically allowing the use of CPVC in single-family residental water distribution systems, for instance. The latter will be a more involved process, and a joining together of several interested citizens to present a fair case might have a better chance of success. In any event, you will doubtless have to study up on the uses of plastic piping and the advantages and disadvantages of its use in residential application. You should point out that plastic plumbing materials are widely accepted throughout the country by a great many local codes, and in general rebut and overcome the arguments of local officialdom. This may not prove easy, but might be well worthwhile.

Homeowners' Rights

There is one other point with regard to plumbing codes, and particularly the National Plumbing Code, that might be of interest to you. In many areas the homeowner is enjoined by local codes from doing his own work. This is a travesty and an infringement upon your rights as a property owner, particularly in the case of a single-family residence. The complete control of the premises should be yours alone, and in any event, a suitable series of inspections can easily be arranged in matters where health and safety might be affected. If there is anything you can do to effectively fight such a stipulation and you feel so inclined, by all means, do so.

Section 1.10 (c) states the following, "Any permit required by this code may be issued to any person to do any plumbing or drainage work regulated by this code in a single-family dwelling used exclusively for living purposes . . . in the event that any such person is the bona fide owner of any such dwelling and that the same are occupied by or designed to be occupied by said owner, provided that said owner shall personally purchase all materials and shall personally perform all labor in connection therewith."

The National Plumbing Code says that you can do your own work in your own home. Common sense says likewise, unless you yourself choose to allow someone else to do the work. There seems little good reason that any local bureaucrat should dictate otherwise.

There is a bottom line to this discussion. If there is a plumbing code in effect in your area, whether local or national, and particularly if it is enforced, you have three alternatives. The first is to ignore the code and just barge ahead; that's not very smart. The second is simply to comply completely with all provisions as necessary. The third is to file an appeal and seek a variance to whatever article or points you disagree with. If you chose to appeal, do so with dignity and intelligence and with an open mind, but never in a bull-headed, obstreperous manner. If the decision goes against you, accept the fact gracefully and swallow your anger. Whether you appeal or not, follow the code provisions faithfully. Treat the plumbing inspector courteously and in general be cheerful and aboveboard about the whole affair. There is an old saying that you can catch a good many more flies with honey than with vinegar. Likewise, you will find it much easier to successfully complete your plumbing installation by working with the various members of officialdom in a relatively pleasant fashion.

SUBDIVISION CONTROLS

More and more often these days as residential subdivisions and developed tracts of single-family homes spring up about the countryside, residents are forming groups armed with written controls to regulate certain aspects of the particular area in which they live. Such controls may come about because of a lack of or particularly weak zoning, or maybe in additon to zoning administered at state, county or municipal level. And though the particular controls, or rules and regulations, vary from development to development and are entirely locally devised and written, and affect specifically only those folks who live in each small area, nonetheless they are effective and enforceable and for the most part incontrovertible. If the rules and regulations are properly written, they are perfectly legal and can be enforced. Such enforcement is usually accomplished at the first level by authority to turn off an individual's water supply.

This is crude but most effective. The power is backed up by recourse to obtaining aid from local law enforcement agencies, filing lines and law suits and by various other courses of legal action.

Control of the Regulations

Control of these rules and regulations is usually handled by a particular group of homeowners who reside within the affected area. These folks may be members of a board of directors of a homeowners association, or some similar group, or perhaps members of an architectural control control committee. The rules and regulations themselves often take the form of a basic set of covenants, which can be added to as time goes on and need arises. There may be instead of or in addition to the convenants a set of rules pertaining to architectural controls ranging from simple and rather elementary to extremely stringent. In any event, every property owner within the affected area of subdivision is legally bound to comply with all of these rules by virtue of having signed his deed or other legal documents pertaining to the conveyance of the property.

More often than not, these rules and regulations have little to do with the interior makeup, design or construction of the residence. Rather, they are concerned with such matters as the proximity of structures to property lines, the visual impact of various appurtenances such as clotheslines, fences, trash containers and such. With respect to plumbing systems the consequence to the individual homeowner is likely to lie in two areas, water supply and waste disposal.

Water Supply and Waste Disposal

There will doubtless be certain procedures to follow in obtaining permission for a water tap, and the location of that tap may have to be determined by those in authority. The process of tapping in a new water supply line will probably have to done by a designated contractor, and an inspection made before the job is finished. You can also count on a fee of some sort, as well as periodic assessments for water use. The installation of a water meter may or may not be required; it if is, the installation may or may not have to be made by a designated contractor.

The rules may also spell out restrictions relative to the placement of a septic tank that govern its proximity to your house, property line, other buildings and the like. The type of septic tank and/or leaching field that you can use may also be specified, or certain kinds disallowed. Or, if the subdivision operates its own private sewage disposal system and plant, there will be a certain body of information regarding proper connection to that. In both cases, you may find that the use of plastic plumbing components is not permitted, or only for one or the other. If so, there is usually a procedure to follow to make an appeal for variance to those in authority, just as there is under zoning laws and with respect to building codes.

If you happen to live in a subdivision or a tract housing area that is governed by a homeowners association or some similar group, investigate what rules and regulations are in operation and get a solid idea of how they might affect your plans. Most likely this will be of interest to a prospective buyer of a home in such an area. There is also a possibility that changes, additions or renovations that you might have in mind as a current homeowner may also be affected by these rules. You can check easily enough simply by locating one of the persons in authority and obtaining all the necessary details.

WATER PROBLEMS

Another area that should be of some concern to a prospective property buyer is the various difficulties that can arise in trying to develop a residential water supply where there are no municipal mains. The alternative to mains, of course, is to draw water from either a well or a surface body. Unfortunately, there are times when all manner of legal complications stand in the way of this seemingly simple process.

For instance, occasionally deed restrictions pop up whereby the previous owner, or an owner several other owners removed, retains the right to all water on or under the parcel of property. Therefore, the parcel is effectively waterless and anyone who takes water from that parcel, whether surface or underground, is actually doing so illegally. A similar situation can arise where a parcel of land is conveyed with only surface rights to the property; all underground rights for minerals and such are reserved to the seller. This may mean in effect that the new owner cannot drill a water well, since he has no rights to anything that lies beneath the surface of the ground. If there is no available surface water, he is stuck.

Even if there is surface water, he may still be stuck. In many areas of the country, most or all of the water that flows in the streams, creeks and rivers is owned by somebody. That water is not available for the taking, but has either been adjudicated for some particular purpose or use or actually belongs to a water company, syndicate or some similar group. The only way an individual homeowner can legally pump water from that source, even if only for domestic use, is to purchase it. And that is usually a complicated if not impossible proposition. Should you be contemplating buying a piece of property in an area where water rights are second only in importance to God, be sure to check very, very carefully before you pump any water from a surface body.

Nor, in many circumstances, is the situation much easier with respect to wells, either dug or drilled. There are plenty of places now where it is no longer possible to just move in a drilling rig and go to work. First you must prove a need for the water. You must present an accurate estimate of how much water will be needed, and you will have to state the exact purposes that the water will be used for. Then you must apply for a well permit, which may or may not be granted. Applications for well permits may be handled and processed at the state level, but you may able to apply at the county or some other level of government. You can check this by contacting the nearest building office. If you contemplate purchasing property in an area where well permits must be issued before any work can be done, you might wish to make the issuance of a proper permit one of the stipulations in the property purchase agreement.

WASTE DISPOSAL REGULATIONS

If you connect your house drain to a municipal sewer system, there are not likely to be any rules and regulations involved except for those pertaining to the actual proper installation of the sewer line and connection at the main tap. Such matters are generally governed by the municipal, or perhaps private, authority that operates the system. However, if you connect to a private septic system, the situation is considerably different.

The rules and regulations that govern the use of septic or other types of private waste disposal systems are likely to be somewhat involved. They are generally formulated at state level, but are actually administered at county or some lower level of local government. The state health department is usually the ultimate authority, with the county health department doing the field work. There are very few areas left in the country where some sort of rules and regulations for private sewage disposal installations are not at least theoretically in effect, in the interest of promoting public health and safety and minimizing any potential hazards from improper disposal.

Septic System Instructions

In most circumstances, the type of disposal system that is installed is the septic tank and leaching field arrangement, and that is our principal point of interest here. Note, however, that there are alternative system as well. The common privy is still in use in many rural areas, and chemical toilets, incinerating toilets or composting toilets are also possibilities. In most cases, these too are affected by state health department rules and regulations where in effect. Details on exactly what is involved and what specifics you must comply with can be obtained by contacting either the country or the state health department.

In general, however, the entire septic system installation must be made according to a specific set of instructions. These instructions vary considerably from place to place and according to local conditions like soil absorption capability. The system must be inspected and approved before being put into operation. In most places, the process starts with the filing of an application for a septic system permit. This may or may not be accompanied by a fee, but usually is. Soil tests may or may not be required, but percolation tests almost certainly will be needed. In some cases these may be done by qualified health department personnel, or at least verified by them.

Approval of the System

The entire system must be planned and drawings made for approval by the authorities. Certain minimum, and perhaps maximum, distances will have to be maintained, such as sewer line lengths, proximity of the septic tank to the house, distances of leaching field from property lines, and safeguarding of water wells or other water sources and supplies. Septic tank construction or type may be specified, and the type of piping and fittings that may be used will probably be defined as well. The leaching field type and size must be worked out according to local requirements in conjunction with the absorption capability of the soil and the number of plumbing fixture units that will be connected to the system.

These, and doubtless other points as well, will be checked on a continual basis by a qualified inspector as the construction job progresses. After the job as been completed, tested and found to be satisfactory in all respects, you will be given a certificate of final approval that will allow you to operate the system.

Chapter 6

Designing the Plumbing System

Designing a complete residential plumbing system, or a subsystem, is really not a terribly difficult chore. This is especially true where the do-it-yourselfer plumber designs his own system, and his notes and drawings will be later followed only by himself. After all, as the homeowner and designer of the system, he can interpret exactly what needs to be done at any given point better than anyone else, and has no need of extensive plans or specifications. Nonetheless, there are a number of essential points that must be considered in the design of the plumbing system. There are also two or three different approaches to the system planning that are particularly useful for the do-it-yourselfer.

DESIGN TO CODE

As has been mentioned previously, in most areas of the country there is a plumbing code in effect, either national or local, that must be followed during the design and installation of a residential plumbing system. Even if there is no code in actual effect, following the major precepts of the National Plumbing Code is not a bad idea at all. So, as you begin to devise and develop your plans for a plumbing system, check through the applicable codes to see which articles in particular might affect your plumbing system and be governed accordingly. Those features that are specifically recommended or made mandatory must be included in your plans, and those procedures, techniques and materials that are code-specified must likewise be made a part of the installation.

Incidentally, in code language, the words "shall be" or "must" always mean that particular item is mandatory. The word "recommended," or its equivalent, means that this is what the authorities would like to see, but you are not bound to follow the suggestion. The interpretation placed upon any particular article is frequently variable depending upon the judgment or whim of the local inspector. In sum, design to the code and you will have fewer difficulties.

SYSTEM DESIGN

The design of the system is the first stage in working up a plumbing installation. The object here is to define just what will be done, and in general, with what. You must work out the course of your water supply line, determine its length, lie and diameter, the point of entrance into the house, the type, size and location of the water pump if there is to be one and similar items. You must determine the plumbing fixtures that will be used and their approximate locations within the building, the number of water distribution outlets that will be needed now and in the future, and the type, size and approximate length of major runs of piping that will be required. You must work out water pressure and flow details so that you will have neither too much nor too little.

The DWV system likewise must be essentially lined out with approximate sizes and lengths of pipeline determined, as well as the pipe type. Approximate locations must be worked out for waste stacks, vent stacks, major branch drainage lines, the house main drain and its point of exit from the building. The length, course and routing of the sewer line need to be determined. If there is to be a septic disposal system installation, the details of that must likewise be worked out.

All of this, you will note, has little to do with specific details of the installation job itself. At this point, you are merely setting the major parameters for the system as a whole. We'll investigate just how this is done in a moment.

SYSTEM PLANS

There are two basic ways that the do-it-yourselfer can approach the business of laying out plans for the installation of a residential plumbing system. The first is the more traditional in the plumbing industry, and that consists of making some sort of plans or diagrams that outline the entire system piece by piece. The second method is probably the more traditional with do-it-yourselfer, and might be called the empirical planning method. This relies to only a limited extent upon plans and notes and much more upon perhaps a sketch or two and direct piece-by-piece installation.

Making Drawings

If you wish, you can make drawings of the entire plumbing system or subsystem that range anywhere in complexity from very simple schematic or symbolic sketches to complete and complex plans. These plans can be drawn up in accordance with current accepted drafting practices and using all of the proper symbols.

Simple schematic sketches are the easiest to make up and, if made by the same person who will do the installing, are probably all that will be necessary. The drawing shows, with the aid of standard symbols or with symbols made up by the draftsman, the principal outline of the piping system, including valves and fittings (Fig. 6-1). In such drawings, many of the minor fittings such as slip nuts on traps are omitted; major fittings and valves are included. The object is to line out all of the major pieces of the system in order to make it easy to work up a material takeoff, which in turn is little more than a grocery list. Often the previous experience and the knowledge that the plumber already has enables him to scan such a drawing, pick off and list the major components, and add from memory all of the little bits and pieces and supplies that he will need to complete the system.

A more complicated drawing will detail every length of pipe, every fitting, every plumbing fixture and every accessory part in the entire system. The drawing may actually be scaled, or the necessary dimensions such as those for pipe lengths may be written in. Very complex drawings may also include the more-or-less exact locations of the pipes within the walls, show the points where they make directional changes or where they are to be supported, and similar minute details. Though the drawings are not three-dimensional, they are so complete that with a bit of imagination they can be made to seem so while being read.

To the neophyte do-it-yourself plumber, however, drawings of this sort are apt to be more confusing than revealing. Not to mention that there is a tremendous amount of work involved in making them. Their advantage is, of course, that nothing whatsoever is left to chance and every minor detail is spelled out; all the plumber need do is follow directions explicitly, and complete the job as ordered. If something happens to be awry, it is not the plumber's fault but the designer's (Fig. 6-1).

The Empirical Approach

The empirical approach is a good method for the do-it-yourselfer who wants to get on with the job at hand. It is practical, time-saving, relatively painless and only requires that the plumber proceed with a certain amount of caution and common sense. The first step, of course, is to work out the design details of the system or subsystem. In other words, decide what kind of pipe you want to use, what sizes you will need, what fixtures you are going to install and such like items.

The next step is to betake oneself to the actual site of the plumbing installation and decide what is going to go where. Spot all of the plumbing fixtures in their appropriate locations, and do so accurately. Then figure out, simply by looking around, just what the best way is to get a pipeline from point A to point B. At the same time, figure out what fittings will be required in order to put in that section of pipeline. Make up a list of parts and measure out the approximate pipeline lengths. Also, you might make a quick sketch of how you plan to make the installation so that you don't forget in the meantime what it was that you were supposed to be doing. Alternatively, make a series of marks and notations directly upon the framing members, floors, walls or whatever. Then go on to the next section of the plumbing system and figure that out.

By the time you get done you will have a parts and supplies list that can be boiled down easily into a shopping list. Then all you have to do is buy the materials and start plugging the system in. It's simple, but effective; just make sure that you don't forget anything or make some serious miscalculation. If any questions or doubts arise as you are trying to work out some particular pipeline section, stop and investigate the situation to get some satisfactory answers. Don't just dive in headfirst. Installation of plastic plumbing system is so simple that you are unlikely to run into many, if any, problems.

COLD WATER SYSTEM DESIGN

Designing a cold water residential plumbing system is only difficult if you don't know how. Knowing how requires only the digestion of a certain amount of basic information; after that the job is painless. The details of system design can indeed become very complex and confusing if one wishes to define everything down to the last decimal point, but for our purposes this is totally unnecessary. It is a virtual impossibility, for a number of reasons, to refine such details as available water pressure at various taps under various given sets of circumstances right down to the nth degree.

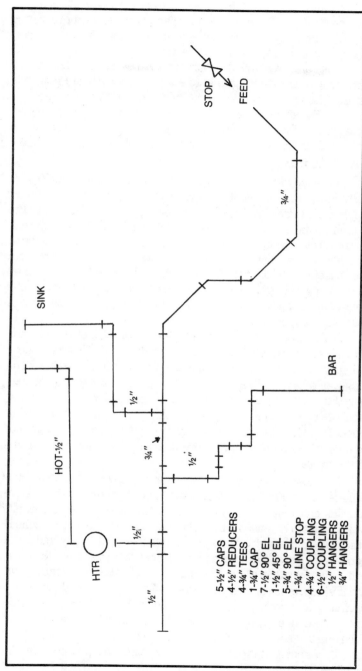

Fig. 6-1. Simple schematic plumbing layout, taken from a plumber's actual sketch of a job. The only change was to straighten lines.

211

The end result of a long series of calculations is nothing more than an approximation anyway. And an approximation is all that is necessary, particularly for residential service.

Therefore, we will deal largely here with relatively broad estimates, rounded-off figures, average or approximate demands and the like in order to keep the picture as clear and simple as possible. By following the same procedure in making your own design calculations, you will readily be able to devise a system that is eminently workable and will give you no trouble.

Basic Parameters

To begin with, here's some background information. You already know that water is delivered to you through your water supply main under pressure. What you may not realize is that the pressure where the water leaves the main and the pressure it enters through your foundation are not the same; the pressure at your house is somewhat lower. As the water travels upward from your basement to the first foor, the pressure is lower yet, and by the time it reaches the end of the line on the second floor it is still lower. Operating water pressure is not equal throughout all parts of your plumbing system.

The reason is simple enough; the inside wall of the pipeline offer a certain amount of resistance to the flow of water. Every fitting, valve or meter in the pipeline adds more resistance. The weight of the water itself in the vertical pipes—the horizontal ones make little difference—must be overcome, which results in further pressure loss. So, enough water pressure at the main to promote a modest but slow flow of water at the first spigot on the line might not be enough to produce much more than a dribble way at the far end of the system.

The quantity of water required at any given time also has a definite effect. For instance, assume water pressure in the system remains constant. Sometimes it does, and sometimes it doesn't, which is another reason why making totally accurate calculations is practically impossible. You open one faucet all the way and get a good water flow. You open another faucet, but this time the flow is somewhat less than at the first faucet, and further more the flow at the first faucet is also lessened somewhat. Open a third faucet and each flow is lessened yet again by a fraction. In theory, you could open up enough faucets on a pipeline to reduce the water flow at each one to a trickle. The reason is that when you increase the flow rate by opening more taps, you decrease the pressure in the pipeline. The pressure available at the main is fixed and there isn't any more; if you split the pie up into enough wedges, there won't be much quantity in any one of them.

Obviously in any residential water distribution system we want to avoid the problems of low pressure and insufficient flow. What we do want is adequate pressure at all the taps and a sufficient flow or volume of water so that we can use a reasonable number of plumbing fixtures at the same time. This is accomplished by providing a sufficient pressure at the main and by installing the correct sizes of pipe throughout the system. At the same time, we want to have a system that combines maximum practical efficiency with the lowest possible cost for both material and installation, meanwhile working up a layout that involves the least amount of time and effort. A few basic simple calculations paves the way for such an installation.

Estimating Water Needs

The first step is to figure out how much water you probably are going to need. There are three aspects to this: the water needed when nothing is turned on (zero), the water that would be needed if everything in the house were turned wide open (more than most supply lines could deliver), and a happy medium that would most likely be used under normal, average conditions at any given time. This is called the *probable demand*, and is the figure we are seeking now. The probable demand is not a fixed or an accurate quantity or volume of water, but rather a reasonable assessment of what perhaps would be needed. It is an educated guess, nothing more, and can't be pinned down any better than that.

Begin by making a list of all of the water outlets or usage points in the entire house. List them individually so that you don't miss any: three washbasins, laundry sink, two kitchen sinks, dishwasher, two toilet tanks, shower, bathtub and so on. Don't forget such items as sill cocks, yard hydrants and the like.

Now, over the years field experimentation, testing and accumulated experience have led to a series of desirable or recommended *flow rates* for various plumbing fixtures. These rate have been found to be satisfactory for normal usage in most situations. As you might expect, they vary somewhat from fixture to fixture; they are also based upon certain pressure minimums in the supply lines. If in any individual case the pressure is greater than these minimums, the flow will also be greater and thus even more effective. Though the desirable flow rate figures vary somewhat, for our purposes here it is reasonable to simply use a blanket figure for each plumbing fixture of 5 gallons per minute. This is a bit for a bathtub, and a bit high for sink or washbasin faucets. Since this is an approximation only, though, no harm will be done. Add up all of the plumbing fixtures and multiply the total by five. If you have 15 fixtures, you will need a flow rate of 5 gallons per minute per fixture or a total of 75 gallons per minute. This is what the system must be capable of delivering if all fixtures are running full blast at the same time, the *total demand*.

Obviously this condition seldom obtains. A good part of the time only one fixture will operate at a time, while several might be turned on at once at infrequent intervals. So, the likelihood that you will need the full 75 gallons per minute is small. Thus, you need to settle upon a reasonable probable demand figure.

Past experience has shown that if there are from 5 to 25 plumbing fixtures in a house, the probability of simultaneous use ranges from 35 to 50 percent. If there are 25 to 50 fixtures, that probability of simultaneous use drops to 25 to 35 percent. Choosing a lower percentage results in a more conservative design and the lesser likelihood of low flow rate resulting from a great number of fixtures being turned on at once. So, to pick a conservative estimate for your 15 fixtures, you might choose to use 45 percent. This means that your total probable demand will be in the area of 45 percent of 75 gallons per minute, or 33.75 gallons per minute. We can round this figure off to 34 gpm, or even 35 to make matters easier. This is what the water supply delivery rate should be, give or take a couple of gpm, for you to be on the safe side. Sometimes the demand might be a bit greater, frequently it will be considerably less, but for the most part 35 gpm will do the job.

The next step is to break some of the figures down a bit to figure out just where this water will be used, how much is needed at each different level, and how much water pressure will be needed to get the job done. Go back to your list of plumbing fixtures and determine which ones will be placed on the various floors of the house. Let's assume that the house consists of basement and two floors. There will be plumbing fixtures on each; for simplicity, we'll say five fixtures per level. Therefore, you will require a flow rate at each floor, in terms of probable demand, of 1/3 of 33.75 gpm, or 11.25 gpm. For ease of figuring, we will round that off to 12 gpm. This translates to a total probable demand entering the house of 36 gpm. That's one gallon more than our previous overall estimate, but that's all right. The difference is slight and either figure can be used.

Water Pressure Requirements

The next consideration is the amount of pressure required to operate each plumbing fixture satisfactorily. This figure varies with different kinds of fixtures. On the low side, a bathtub faucet will function fine with a minimum pressure of 5 pounds per square inch water pressure. A shower, on the other hand, requires 10 psi. Virtually all standard household fixtures operate satisfactorily at pressures within this 5 to 10 psi range, with one exception. A sill cock with 50 feet of garden hose attached requires a minimum pressure of 30 psi to achieve its desirable flow rate of 5 gpm. Leaving this exception aside as a noncritical factor, we can assume that a water pressure of 10 psi, as a rule of thumb, is adequate and in some cases more than enough to satisfactorily operate the household plumbing fixtures. So, that's the minimum pressure needed at each outlet. Note, though, that much more satisfactory operation is achieved with higher pressures, and in fact most household plumbing fixtures do operate at higher pressures.

Top Floor Pressure Loss

The next step is to determine what the water pressure must be at the main to overcome the various pressure losses along the line and still achieve that minimum 10 psi at the far end of the system. The easiest way to do this is in steps, starting with the top floor. For this you need to know within approximately a foot, where each plumbing fixture will be located. Start at the point where the supply riser, the vertical feed pipe coming up from the first floor, joins the lateral, or horizontal, pipeline that will feed the top floor fixtures. Measure the whole length of the pipeline from one end to the other, following all of the jogs and turns that the line might make. This process involves only the main supply line that will pass close by the plumbing fixtures; do not include any branch lines or the short lengths of pipe or tube needed to connect the fixtures to the supply line. As you go along, count up the number of fittings in the line.

Each foot of piping represents a certain amount of friction loss, or resistance to the flow of water, that in turn leads to a certain pressure drop. The farther the water must travel, the greater is the drop. Also, each fitting represents a certain amount of flow resistance, and hence an added pressure loss. There are tables available that are frequently used by professional plumbers and plumbing systems designers which list the exact amount of pressure loss for every conceivable type of fitting and every different kind of pipe that

Table 6-1. Fitting/pipe pressure loss equivalencies.

FITTING SIZE, INCHES	PIPE EQUIV., FEET
3/8	1/2
1/2	1
3/4	1 1/4
1	1 1/2
1 1/4	2

might be used. So, the pressure loss in a give pipeline can be computed quite closely for any given input pressure.

However, for our purposes this procedure unnecessarily complicates our calculations, since at best we will only end up with an approximation anyway. Tables list the fitting pressure loss in terms of pipe diameters' equivalency, or equivalent length of pipe in feet. We will use the latter since it is simple. Furthermore, we will use a rule of thumb that proves out perfectly nicely for such applications. Consider that each fitting in the line is the approximate equivalent of a 90-degree elbow, whether it actually is an ell or not. The exception to this rule of thumb, incidentally, involves the inclusion of valves in the supply line; a gate valve causes considerably less friction loss than a 90-degree elbow, while a globe valve occasions about eight times as much. We will assume for purposes of illustration that there are no main-line valves involved, since there almost never are in the distribution side of a residential water supply system.

In Table 6-1 you will find the necessary figures. Let's assume that the length of your top floor piping turned out to be 35 feet, and you plan on using nine fittings. To start of with, we will further assume that you would like to use ½-inch pipe for the main feed line. By making a trial analysis, we can determine if that size will be satisfactory, if you must go to a larger size, or if you can perhaps economize by using a smaller size. From the table, you can see that each fitting is equal to about an extra ½-foot of pipeline. Therefore, the nine fittings are worth an extra 4½ feet of pipe, which added to the original 35 feet equals 39 ½ feet of effective pipeline length. Round this off to an even 40 feet for convenience.

Plumbing Fixture Location

Next, consider the plumbing fixture arrangement along the pipeline. If all the fixtures are at the far end of the line, all of the

216

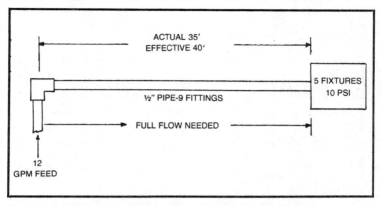

Fig. 6-2. Top floor piping representation; fixtures at end of line.

needed water will flow through the entire pipeline and all fittings to reach its destination, and the whole 40 feet should serve as a basis for calculation (Fig. 6-2). If four of the fixtures are in the middle and the fifth at the far end, the situation is different. Particularly if the end-of-the-line fixture is a low-volume one, most of the water will be used at the midpoint of the pipeline (Fig. 6-4). A conservative estimate might be an 80 percent flow in the first half, or 20 feet of the line.

So, you can make another approximation here, and assume that most of the water will only flow through the first half of the line. Instead of using the full 40 feet of pipeline, we can use just half, 20 feet. Similarly, if the first four fixtures were spread along 30 feet of line, we would use that figure. In other words , adjust the effective pipeline length with regard to the plumbing fixture locations.

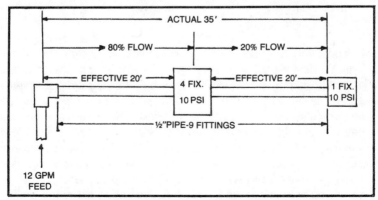

Fig. 6-3. Top floor piping representation; fixtures split.

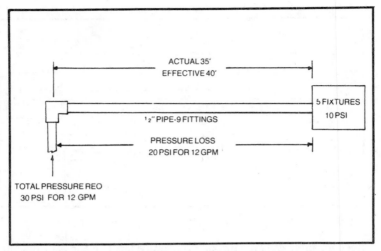

Fig. 6-4. Top floor piping representation showing pressure requirement.

Pressure Loss Computations

To finish the calculations, we already know that we have to have a minimum flow of 12 gpm through this pipeline in order to adequately feed the plumbing fixtures. And a glance at Table 6-2 shows that 10 gpm is the highest flow listed for ½-inch line. Does this mean that ½-inch pipe cannot be used for flows of over 10 gpm? Not necessarily, but at that level a practical limit is rapidly being approached. And this brings up two more points worth of mention. First, this table is compiled of approximate figures only. They may or may not be exactly correct, or nearly so, for any one particular kind of pipe; some types and brands of pipe may carry lower figures, others higher. The best way to make accurate, or at least more accurate, calculations is to use pressure loss tables applicable to the particular kind of pipe you are using.

Second, note that in neither this nor any similar table can you combine flow rates listed in order to arrive at an accurate pressure loss. For instance, if we add a 2 gpm flow to a 10 gpm flow to arrive at our 12 gpm flow in a ½-inch line, the total pressure loss would appear to be 50 psi. But this figure is so far off base as to be unworkable; note the great jump in pressure loss between a 5 gpm and a 10 gpm flow. As water flow increases pressure loss does so at a much greater rate. Adding together a 1 gpm and a 2 gpm pressure loss in ½-inch pipe totals 4 psi, but, as Table 6-2 indicates, a 3 gpm flow results in a 6 psi pressure loss. By the same token, an odd flow rate lying between listed figures must be extrapolated toward the high side. A 7½ gpm flow in ½-inch pipe will not lie halfway between the figures for 5 and 10 gpm, or 30.5 psi, but rather closer to 40.

To get back to the example, now we must determine what to do with our nearly-overloaded ½-inch pipeline. It is now obvious that a 12 gpm flow in the line might result in unacceptable pressure loss, at a guess, maybe 60 psi or so. One solution is obvious—switch to ¾-inch pipe. But that is more expensive, more difficult to work with. Another possibility is to make a few more assumptions and estimates, and stick with the ½-inch size just to see what the results will be. We can assume that our required 12 gpm flow may actually be closer to 10, or even less most of the time. We can further say that even if the flow does reach 12 gpm, the pressure loss will not be bothersome to the occasional users. This leads us to a further guess that the pressure loss perhaps will seldom reach the listed 10 gpm pressure loss figure of 47.0 psi. We can arbitrarily choose that or a somewhat higher figure, say, 50 psi.

If the pressure loss in 100 feet of ½-inch pipe is 50 psi, and for the sake of example we choose the entire 40-foot supply line, the pressure loss will be 40/100 of 50, which amounts to 20 psi. If the pipeline were only 20 feet long, the pressure loss would only be 20/100 of 50, or 10 psi.

Head Pressure

Now let's put this together. Assuming the full 40-foot pipeline, the pressure loss is 20 psi. As discussed earlier, we also need 10 psi

Table 6-2. Approximate pressure loss in pipeline due to friction.

FLOW GPM	PRESSURE LOSS PER 100FT. PIPE SIZE—TRADE				
	⅜	½	¾	1	1 ¼
1	2.5	1.0	0.25		
2	8.5	3.0	0.5	0.25	
3	17.5	6.0	1.0	0.35	0.1
4	29.0	9.5	2.0	0.5	0.25
5	42.0	14.0	3.0	0.75	0.35
10		47.0	8.5	2.5	1.0
15			18.0	5.0	2.0
20			29.0	8.5	3.5
25			48.0	12.0	5.0
30				17.0	6.5
35				26.0	8.5
40					11.0

as a minimum requirement to operate the fixtures properly. Thus, the pressure required at the inlet end of the top floor supply line is 30 psi (Fig. 6-4). The next question is, can we get 30 psi at the top floor? Maybe, and maybe not. That depends upon three principal factors; the available pressure at the main, how far the water must travel to reach the top floor, and how high the water column must rise in the process. Let's see how that works out. At this point, only a rough idea is needed, so we can consider just the *head pressure*, that pressure required to lift the weight of the water.

First, determine the depth of the water main below the street. Then determine the elevation of your shutoff valve within the house above or below street level, and add to or subtract from the main pipeline depth as necessary. To this, add the vertical distance from the shutoff valve to the point where the riser supply line feeds the top floor horizontal supply line. Let's assume that the main is 8 feet underground, your shutoff valve is 5 feet above street level, and the distance from the shutoff valve to the top-floor joint is 18 feet. This totals 31 feet (Fig. 6-5). By consulting the water pressure conversion in Table 6-3 you can see that a 30-foot head requires a pressure of 12.99 and a 1-foot head requires a pressure of 0.43, for a total of almost 13½ psi to get the water up to the top floor.

To this must be added the 20 psi loss in the top floor pipeline, and the 10 psi required for the fixtures. This is a total of 43½ psi required at the main. So far we have not considered frictional losses in the supply line, the riser line and the various fittings that will be required, not to mention the water meter. Unless the pressure in

Table 6-3. Head pressure loss in pipelines.

HEAD (FEET)	PRESSURE (PSI)
1	0.43
2	0.87
3	1.30
4	1.73
5	2.17
6	2.60
7	3.03
8	3.46
9	3.90
10	4.33
20	8.66
30	12.99

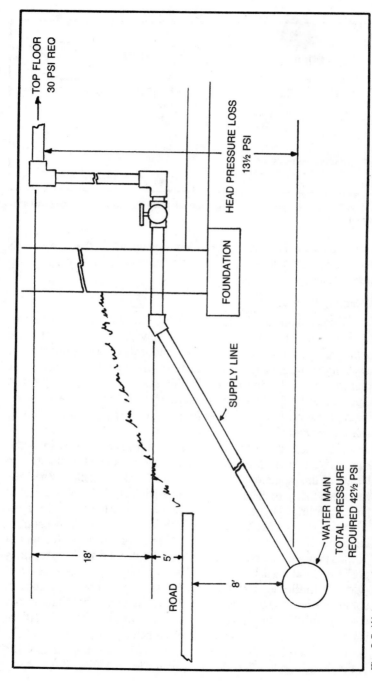

Fig. 6-5. Water supply piping representation showing partial pressure requirement to feed top floor.

TOP FLOOR
30 PSI REQ

HEAD PRESSURE LOSS
13½ PSI

FOUNDATION

SUPPLY LINE

WATER MAIN
TOTAL PRESSURE
REQUIRED 42½ PSI

ROAD

18'

5'

8'

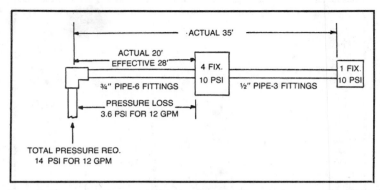

Fig. 6-6. Top floor piping representation showing pressure losses and requirements.

the water main is exceptionally high, it is apparent that the ½-inch top floor supply line size may be inadequate, or perhaps even unworkable.

Changing Pipe Sizes

Does this mean that some of the plumbing fixtures will have to be removed or moved closer to the supply riser outlet? That's one solution, of course, but usually an unnecessary one. You can't do anything about the 10 psi required for the fixtures, but you can about that 20-pound loss in the supply line. Remember, the bigger the pipe diameter, the less the pressure loss for a given flow rate and input pressure. Let's see whether going up one pipe size to ¾-inch will be of any help.

Assume that most of the water use will be at the midpoint of the run, and that there are six fittings needed to get to this point. The length of the pipeline is 20 feet, and by checking Table 6-2 we see that each fitting in ¾-inch size is the equivalent of 1¼ extra feet of pipe per fitting. This give us a total effective pipe length of 27½ feet, which we can round off to 28. Again checking Table 6-2, by estimating between the lines we see that the pressure loss per 100 feet of ¾-inch pipe for a 12 gpm flow (the amount needed at the fixtures) is perhaps about 13 psi. Thus, 28/100 of 13 equals just over 3.6 psi. Add this to the necessary fixture pressure of 10 psi for a total of somewhat less than 14 psi required at the top floor distribution line, considerably less than that required by ½-inch pipe. Note that there is no need to make the entire run of ¾-inch pipe; the second half of the line can be ½-inch (or maybe even smaller, depending upon the fixture) with no ill effects and a savings in cost (Fig. 6-6).

Top Floor Riser Pressure Loss

Now let's find out what we need for a riser pipe and what the pressure losses will be there. Obviously if the first section of the top floor supply line is to be ¾-inch diameter pipe, a smaller size cannot be used for a riser, though a larger one could be. The ¾-inch riser will run from a tee at the first floor level to a point approximately at the second floor level; let's assume this distance to be 8 feet (Fig. 6-7). We need a 12 gpm flow to the second foor, and we have already determined that the pressure loss per 100 feet of ¾-inch pipe is 13 psi. The pressure loss in only 8 feet of pipe, therefore would be 8/100 of 13, or a bit over a pound.

Since this is a vertical rise, we must also consider the head pressure loss. By consulting Table 6-3 we can see that this amounts to 3.46 psi. If we round both figures off and add them together, we have a total pressure loss in the riser of about 5 psi. By adding this to the pressure requirements at the second floor of 14 psi, we see that a total of 19 psi will be needed at the first floor tee to provide an adequate upward flow.

First Floor Pressure Loss

Now, how about the first floor? Calculations for the requirements for the first floor supply line are made in the same way as in

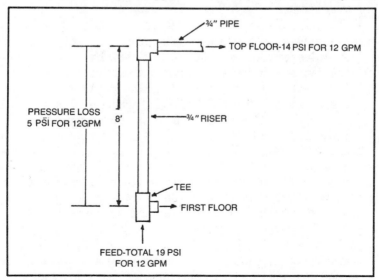

Fig. 6-7. Top floor riser piping representation showing pressure losses and requirements.

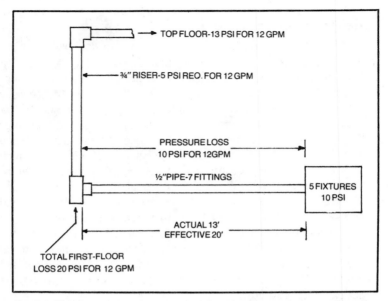

Fig. 6-8. First floor piping representation showing pressure losses and requirements.

those for the top floor. We can dispense with the explanations and carry on with the calculations. There are five plumbing fixtures here, requiring a 12 gpm flow. All five fixtures are closely grouped, we'll assume, and so can be treated as a unit. The pipeline distance from first floor tee to fixture grouping we'll arbitrarily take to be 13 feet, with seven fittings in the line (Fig. 6-8). By making the first calculation for ½-inch pipe, we see that the fittings are the equivalent of an extra 7 feet of pipe, for a total effective run of 20 feet. The pressure loss for 100 feet of ½-inch pipe roughly is 50 psi (assumed earlier). Since this line is only 20 feet long, the pressure loss is the equivalent of 20/100 of 50 or 10 psi. That is not an unreasonable figure, and we know from the top floor calculations that if need be we can substitute ¾-inch for a substantial reduction in pressure losses. For the moment, let's carry on with the 10 psi figure. The required fixture pressure, just as on the top floor, is 10 psi. Adding this together, we find that the required first floor pressure is 20 psi.

Note that there is not much difference between the required pressures at the first floor tee for the two floors—the first floor is 1 pound higher. The only effect is that the pressure available for the second floor will be a bit greater than originally calculated, and is merely a bonus. As long as the first floor pressure is greater than the top floor, all is well. In fact, if the first floor pressure requirement was less by 2 or 3 pounds, or probably even 5, still no harm would be done and the system would likely be perfectly functional. If there was a substantially lower pressure requirement at the first floor, obviously the higher pressure should be met, at least approximately.

First Floor Riser Pressure Loss

The riser to feed the first floor from the basement must be of ¾-inch size minimum, since it will also feed the riser from the first to the top floor and cannot be smaller than that pipe. Making calculations in the same manner as for the first-to-second-floor riser, if the head is 8 feet as before, the head pressure loss will be approximately 4 pounds. Let's assume in this case that the riser must take a slight jog and is 10 feet long including two fittings equal to 2½ feet of pipe (Fig. 6-9). We need a water flow of 12 gpm to the top floor and another 12 gpm to the first floor, for a total of 24 gpm. Consulting Table 6-2, we see that a 24 gpm flow through 100 feet of ¾-inch pipe results in a pressure loss of approximately 42 psi. Our riser is only 1/10 of this length, so the friction pressure loss will amount to about

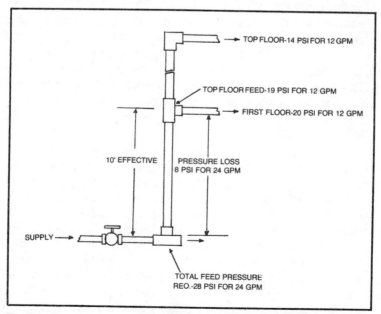

Fig. 6-9. First floor riser piping representation showing pressure losses and requirements.

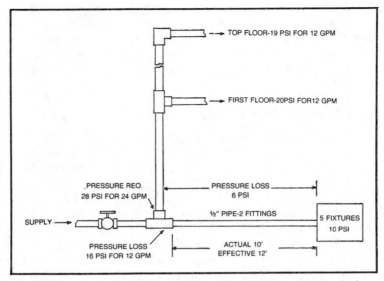

Fig. 6-10. Basement piping representation showing pressure losses and requirements.

4 pounds, rounded off. The total loss in the riser, then, is 8 pounds from the main shutoff valve to the first floor tee. The total pressure required to feed the first floor, then, is equal to the required first floor pressure of 20 psi plus the riser pressure loss of 8 psi, for a total of 28 psi. This is a higher pressure than required by either the first or the second floor branches in themselves, so we are in good shape as long as the higher pressure requirement is met.

Basement Pressure Loss

Now on to the basement. Again we have five plumbing fixtures to consider with a minimum psi requirement of 10 pounds and a gpm rate of 12. We can assume that all five fixtures are grouped within 10 feet of the main shut off valve, with only two fittings. Using ½-inch pipe size, this makes for an effective pipeline length of 12 feet (Fig. 6-10). Using an approximate pressure loss figure of 50 psi per 100 feet of ½-inch piping, we can figure that 12/100 of 50 equals a pressure loss from friction of 6 psi. Added to the fixture pressure requirement of 10 pounds, we have a total of 16 psi needed for the basement piping. Obviously, since we need 28 pounds to supply the first floor and the basement pipeline connection will be made at approximately the same point, we have more pressure than necessary once again. Everything should work out nicely.

Supply Line Pressure Loss

Now all that is left to do is work out the details for the water supply line. We require a flow rate of 12 gpm for each floor, or roughly 35 gpm total for easy figuring. We also must have a minimum water pressure of 28 pounds entering the distribution system to properly supply the plumbing fixtures. We will assume that the distance the pipeline must travel from the water main to the main shutoff valve in the house is 50 feet, and that the water must rise vertically from the main to the valve a total distance of 6 feet. We will assume, too, that the water must pass through a stop valve at the main, a curb stop valve at some point outside the house, and the main shutoff valve itself. This water valve, incidently, should always be a gate valve which has low restriction, rather than a globe valve that greatly impedes water flow. Also, the water must pass through a meter at some point. And since ¾-inch pipe size is often specified for water tap and house supply lines, we will use that size in our calculations (Fig. 6-11).

The first step is to find the effective pipeline length. The presence of a gate valve in a ¾-inch line is the equivalent of an extra ¼-foot of pipe. We have three such valves, so we can round this off to an extra 1 foot of pipeline. Checking Table 6-3, we find that the pressure loss in 100 feet of ¾-inch pipe is not listed; the size is not adequate for normal uses. We can prove that by completing the example. The pressure loss might be as high as 100 psi, maybe more. So, 51/100 of 100 equals 51 psi. To this we must add head pressure loss; according to Table 6-3, this is 2.6 psi which we will round to 3. So far, we have a total of 54 psi. That leaves just the meter.

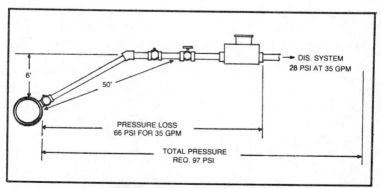

Fig. 6-11. Water supply piping representation showing pressure losses and requirements to feed total domestic system.

The pressure loss is variable through a water meter, depending upon flow and also the particular construction and characteristics of the meter. The best bet is to obtain specific information for the particular application. In this case, the pressure drop could be as much as 15 psi. That makes a total of 66 psi required at the main to supply water to distribute water throughout the house, for a grand total of 97 psi.

All of this is hypothetical, of course, but it does point out the fact that a ¾-inch supply line is impractical in this situation. It is possible to provide a water main pressure of 97 psi, and many run a good deal higher than that. But 80-90 psi is considered quite high for a residential system, even though a few do operate at 100 psi or so. Troubles, though, are bound to arise. This system would be functional at a reduced water main pressure of 80-90 psi, but at the expense of a total flow of less than 35 gpm. Recalculating on the basis of a total flow of 25 gpm, for instance, shows a total pressure requirement of 56 psi required at the water main, a perfectly workable arrangement. Obviously, juggling of estimates and figures along with pressure/gallonage tradeoffs can lead one to a fully workable design.

Making Adjustments

If the available water main pressure happens to be less than the final figure arrived at in your own design, there are several options you can choose. If the difference is marginal, you might elect to live with the situation. All of your figures are merely reasonable estimates anyway, and the difference of a few pounds will seldom if ever be noticeable during actual operation of the water system.

If the available pressure is considerably less than your design calls for, one segment or another of the system will have to be recalculated and redesigned. For instance, one might opt for a larger supply line. The pressure loss at a 35 gpm flow in a 1-inch 50-foot line, for instance, is only about 26 psi. Add the head pressure loss of 3 pounds for 29 psi. A water meter in this situation will cause a pressure drop of only about 10 psi, which brings us up to 39 psi. Add this to the estimated house pressure of 28 psi, and the requirement at the water main is a reasonable 67 psi, compared to the high 97 psi in our example.

The riser pipe from the cellar to the first floor could be figured on the basis of a 1-inch line instead of ¾-inch. Alternatively, one might go back through the entire set of calculations to reassess flow demand, refine the numbers to lower stages, do less rounding off and loose estimating, and come up with a tighter and perhaps more realistic pressure requirement for the supply at the entrance to the distribution system.

For the owner of a private well, the situation is somewhat different. If the well is already in a particular location with respect to the house, obviously the homeowner must live with that situation. If the well is yet to be drilled, it could possibly be positioned just a few feet outside the house and in line with the most convenient position for a main shutoff valve or a pump location. This not only cuts down on piping costs, but also has a beneficial effect upon pressure requirements. Furthermore, the pump can be sized to deliver whatever water flow rate is desired and whatever water pressure is required for the house distribution system (within reason, of course). Instead of having to live with a certain set water pressure at the source, it can be adjusted to suit.

Other Approaches to System Design

The example we have used here is a fairly simple and straightforward one for ease of illustration and calculation. Doubtless your own system will bear little resemblance, but nonetheless this is a good way to approach the design of a plumbing system. The example also points up a couple of important facts. One is that if you simply go ahead and plug into numerous pipes strung all over the building with no regard for pressure drop and the like, you may well end up with a water distribution system that is only partly effective and may not operate as satisfactorily as you'd like. On the other hand, of course, the system might function perfectly well. Many such installations do, simply by sheer luck and the circumstance that the available water main pressure happens to be fairly high.

The other fact is that by condensing and compressing the entire plumbing system as much as possible, and by using the smallest practical pipe sizes and a minimum number of fittings, you can construct a simple plumbing system that will work effectively even with low or fluctuating pressures at an absolute minimum of expense. The stacked plumbing system, where plumbing fixtures are located back-to-back, directly adjacent to one another and/or directly above one another, makes for a much simpler and more efficient plumbing system, with respect to both water distribution and DWV lines. Making the calculations is easier, too.

Should the would-be home plumber go to such lengths in making calculations for a new cold water distribution system? For the best results, the most efficient and economical system, and the best realization and knowledge of what the system is and how it will work, the answer must be yes. From a practical standpoint, however, one could also say no, not necessarily. In fact, successful residential plumbing systems can be, and have been, installed on an experimental basis by following one's own proven experience or that of someone else. What has worked well previously will doubtless work again with the same results, provided that the installation conditions are approximately the same as well.

Thus, in designing and/or installing a particular plumbing system or subsystem, one might draw on one's own fund of experience from past jobs. Consult with an experienced plumber about the various details of the job. Also follow approved practices and tables as presented in literature on the subject, or simply see how the next-door neighbors' plumbing systems are installed and follow suit.

To choose a random example, one might first determine the available water pressure at the supply source. If that were relatively high, let's say 75 pounds or so, one might logically assume this to be sufficient for an ordinary, average cold water distribution system. To minimize any pressure drops, the supply could be brought to the house in large pipe, perhaps 1-inch or even 1¼-inch. Risers could be kept as short as possible in order to minimize head pressure, and installed in ¾-inch or 1-inch size to minimize pressure loss. For the same reason, fixtures could be grouped, pipelines kept short and a minimum number of fittings used. This also saves money. Feed lines to major plumbing fixture groupings could be installed of ¾-inch pipe size, with short branch lines of ½-inch pipe size. The branches to and stub-outs for plumbing fixtures could be made no less than 2 feet.

Under these circumstances, a fairly compact water supply and distribution system serving an average two or three bedroom home of typical design would undoubtedly function well and present no problems. And, in the event that such a problem as low pressure or low volume might occur, further steps could be taken to correct them, though at added expense and labor. In short, this practice, though not a recommended one for guaranteed best results and most economical fabrication of the plumbing system, does work.

Additional Considerations

The next problem is to determine what to use for the cold water supply and distribution system. As to the supply line polyethelene (PE) pipe is a good choice, inexpensive and easy to lay because of its flexibility. Very few fittings are required, an important fact for underground pipelines that can not be easily repaired. Polyethelene is widely used for well-and-pump applications, and is also a popular choice for water main tap-ins. In the latter case, galvanized steel or Type K copper tubing may be used to tap out of the main where the pipeline must go under a roadway. Then a transition is made to the plastic pipe. Another common choice for water supply lines is rigid PVC piping. Though this is a bit more work to install and consequently a bit more expensive, it is equally satisfactory in operation.

Certain other parts will also be needed for the water supply line. These are somewhat variable depending upon circumstances. In a water-main tap-in there will probably be a saddle and a stop cock located at the main (these may or may not be provided), a curb cock with curb box and cover located somewhere along the line, and a main shutoff valve located just inside the house. A water meter may or may not be required, and a meter stop cock may or may not be installed.

At some point at or around the main shutoff valve and/or the water meter, a transition is made to a different type of pipe. The most commonly used kind for the cold water distribution system is CPVC. In theory, PVC can also be used, but seldom is because it does not have the capability of carrying hot water. It is simply easier to use one kind of pipe to do all of the distribution piping within the house. A newcomer to the field that will see more and more use as time goes on is PB tubing, an extremely flexible and versatile material. There is no reason why both types cannot be used in the same distribution system. The tubing can be employed wherever its flexibility and long, unbroken lengths are advantageous. The rigid CPVC can be used wherever its characteristics are more desirable for particular installation situations.

Along with the necessary lengths of pipe itself for the cold water distribution system, an assortment of fittings will also be needed. These, of course, depend entirely upon the specific routes, along with the various directional changes that the risers, laterals and branches will take. By and large, a collection of 90-degree elbows, 45-degree elbows, couplings and tees, and perhaps some caps and reducing bushings, is about all that is required. Each stub-out for a plumbing fixture connection should be fitted with a stop valve, either straight or angled as the situation demands. In certain circumstances, the installation of one or two riser or main feed line stop valves might also be advantageous, in order to cut off full sections of the distribution system while leaving others operational. Also, in some installations certain transition or adapter fittings might be needed. All of the distribution system while leaving others operational. Also, in some installations certain transition or adapter fittings might be needed. All of these fittings, of course, should match the piping in both size and material compatibility. Finding out exactly what you need is merely a matter of going through the entire distribution system by sections and making a list of the necessary bits and pieces.

System Layout

As to the layout of the system, let common sense be your guide. Pick the shortest and most direct routes from point to point for the pipelines, using a minimum of fittings. Keep the pipelines out of exterior walls wherever possible, and run them only where they will be exposed to normal indoor temperatures, well above freezing. If pipes must be run on exterior walls, always place them to the inside of the thermal insulation. (In climates where the temperature never approaches the freezing mark, of course, pipe position is of little consequence). Keep all plastic pipelines well away from sources of heat, such as comfort heating radiation units, furnaces, chimneys and smoke pipes, and even refrigerator and freezer coils.

Where hidden connections must be made to plumbing fixtures, as is the case with some bathtubs, arrange the pipelines so that you can build a convenient access hatch over the connections. Make sure that all pipelines are adequately supported, with space allowances for expansion and contraction. They should be set in such a way there is little or no danger of them being punctured by a nail as interior finish materials are put up. Remember to install plastic boilder drains at a convenient spot for an automatic washer hook up, whether you have one now or not, and to provide at least one outdoor sill cock or hose bib. In cold weather climates, these should be of the freeze-proof variety, but in warm climates the plastic boiler drain fitting serves well.

HOT WATER SYSTEM DETAILS

In many respects, the design for the hot water distribution system is a replication of the cold water system. Layouts are figured in much the same way. Determining pipe sizes and pressure losses is done by following the same calculation system as for cold water. There is no main supply line to be concerned with here, of course, but there are riser and lateral feed lines, as well as the plumbing fixtures themselves to consider. In most cases, however, it is entirely unnecessary to go through the whole routine again for a domestic hot water system (so called to differentiate it from heating hot water systems). With a few exceptions, all of the normal household plumbing fixtures that require cold water also require hot water. The exceptions, all of the normal household plumbing exceptions are the toilet tank, refrigerator ice-maker, outdoor hose bib or hydrant, and possibly a sink for a wet-bar that only takes cold water. Also, a dishwasher installation requires only hot water—no cold.

So, from the point of origin of the heated water to most of the plumbing fixtures in the house, the hot water distribution line simply parallels the cold water distribution line. The same pipe sizes can be used, and for the most part the fittings are also identical. Pipe lengths will differ somewhat because of the spacing between the two sets of lines. There is also a possible exception with respect to pipe sizes. Sometimes the next size down can be used where the fixture will require only a modest amount of hot water by comparison with cold. You can determine for yourself where it might be possible to economize a bit with smaller pipe sizes in your hot water distribution system by making the necessary calculations based upon desired flow rate at the fixture and the pressure losses in the feed line. Whatever pipe size strikes an appropriate balance between the two is the size to use.

Determining Hot Water Needs

There are considerations with respect to hot water distribution systems, however, that do not occur in cold water systems. One of the most important of these is determining the hot water demand. You will recall that in making calculations for the cold water system the basis was an assumed probable demand. That same demand stands for the hot water system, insofar as correct pipe sizing is concerned. By using the same pipe size for hot water as for cold, perhaps modified to a smaller size in places as indicated by the results of further calculations, we are assured of the proper flow rate for the hot water. But this is a flow rate in terms of gallons per minute, and to heat that much water on a constant basis is obviously ridiculous. No homeowner could hope to afford either the necessary equipment or the cost of operation. So, we must estimate the probable demand for hot water on a somewhat different basis, using gallons needed over a period of time, usually a 16-hour day or thereabouts.

This has to be broken down into two parts. The first step is to calculate the total number of gallons of hot water you are likely to

Table 6-4. Approximate hot water requirements for fixtures.

Fixture	GAL. REQ.
Tub	20-40
Shower, per min.	3
Lavatory	1½-3
Kit. sink, per meal	2-4
Dishwasher, one cycle	3-8
Clothes washer, one cycle hot	30-40
Clothes washer, one cycle warm	10-20

need during a normal day. This must be an estimate, of course, and Table 6-4 can serve as a guide. Note that the numbers are themselves only estimates. Such figures depend not only upon the type of equipment in use, but also the personal habits of the users. There is no way to pin down exact figures. Cast your mind over the normal daily household routine, and tally up a reasonable total of hot water consumption.

Next, make a few notes as to when this hot water will be needed during the day (Fig. 6-12). This is an important factor, because it will determine in part what kind and size of water heater will best serve your needs. When you draw hot water from a heater tank, cold water enters to replace the hot being used. A certain amount of time is needed for this cold water to in turn become hot and ready for use. Thus, if you install a 35-gallon tank and require 30 gallons of hot water at one point during the day and another 30 gallons 6 hours later, you are in good shape provided that the tank is capable of reheating the water in less than a 6-hour period of time. If you attempt to draw 30 gallons from the tank now and another 30 gallons in half an hour, you will get cold water.

Water Heater Types

Once you have determined about how much hot water you will need for a day, and also what quantities you will need when, you can go about selecting a water heater. There are several kinds from which to choose. The object is to find the smallest size that will adquately serve your needs at a reasonable initial cost and the lowest possible operating cost. Remember, in many households the seemingly simple matter of heating water for general consumption accounts for as much as one-third of the total household fuel expenses.

If your comfort heating system is operated by a steam or a hot water boiler, you can consider using a tankless water heater. There are two types. The direct heater consists of a coil of copper tubing immersed directly in the boiler water. The indirect type consists of copper pipe coiled around the main heating pipe from the boiler. In either case, cold water enters at one end, picks up heat by transfer from the boiler water (there is no direct contact of boiler and domestic water) and goes out the other end as hot water in a continuous process. Because of its low efficiency, the indirect type is not often used. The direct type has a high rate of heat transfer and is a reliable, efficient system. Either type requires, first, the presence of a boiler; and second, that the boiler be run all through the summer months.

The tank-type heater is nowadays the most popular. This consists of a specially designed tank engineered to contain potable water, and equipped with a heating source such as electric elements or a gas burner. Capacities of from 10 to 120 gallons are available in a wide range of tank sizes and shapes. Some are round, upright and

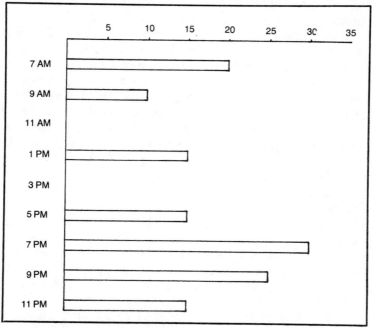

Fig. 6-12. Daily hot water demand.

meant to be placed in a utility room, or in a basement. Others are quite small and can be installed within a kitchen base cabinet. Some are designed to look like cabinets themselves and are freestanding.

Each type of tank is relatively efficient, though there are several designs of each and essential design differences between each type. Whether gas or electric is the best bet for you depends upon several circumstances. One is availability in the size and capacity that you need; another lies in operating costs. An electric water heater also has the advantage of being flameless. The unit can be placed in any handy spot, requires no flue or vent and can be positioned next to combustible surfaces with no hazard. Gas heaters, on the other hand, are somewhat limited in those respects. Another point to consider is the ready availability of the two kinds of fuels. If no natural or bottled gas is currently piped into your house, probably an electric tank would be the better choice. On the other hand, if you have a small and relatively heavily loaded electrical service entrance, and gas is readily available to you, that might be the better choice.

Selecting Heater Tank Size

In choosing a tank size, consider first your total daily requirements for hot water gallonage. If your total is 75 gallons, it would appear at first glance that you might need a 75-gallon tank, and the closest standard size to this is an 82-gallon tank. But now consider the times at which the hot water will be used. If the entire 75 gallons will be consumed within an hour's time each evening, then you do indeed need the full capacity. Suppose that you need 35 gallons in the morning and 45 gallons in the evening. Since there is more than ample time for a tank full of water to heat up to operating temperature throughout the day, you need only provide yourself with the larger of the two peak demand quantities, or 45 gallons. The nearest standard size tank to this is the 52-gallon capacity.

But suppose, on the other hand, that you need several bursts of hot water at various periods throughout the day. Now another factor comes into play, a tank capability known as *recovery rate*. This is the time period required for the tank to heat a full load of water to its set operating temperature from a cold start. Some tanks do this rather slowly, while others are especially designed for rapid recovery. The term, however, means little; what you need to know is how long it will take for the tank to raise the temperature of its full capacity of water by 100 degrees. To turn that around a bit, you must know how many gallons of water per hour will be raised in temperature by 100 degrees.

Here's how it works. Suppose you have a 40-gallon tank heater, and you require 35 gallons of hot water now and another 35 gallons within 45 minutes. If the recovery rate of the tank is such that 50 gallons of water can be raised in temperature by 100 degrees within an hour, you will have sufficient hot water capacity. By the time you get around to using the second 35-gallon batch, the tank should have just about recovered sufficiently to deliver the water. However, if the tank can only reheat 30 gallons of water in an hour, you will likely have an insufficient supply. In that case, you would need either a tank of similar capacity but a faster recovery time, or a tank of larger capacity. The large tank would still have hot water left in it after you used the first 35-gallon batch. It would heat up sufficient extra water to be able to easily provide the second 35-gallon batch.

Choosing a tank with the proper recovery rate is a tricky bit of business that requires you to make a fairly accurate assessment of your hot water needs. If you buy one that is too small or has too low a recovery rate, you will not have an adequate hot water supply and may have to resort to installing a second tank in tandem or in a different location. Incidentally, in large, sprawling houses with two or more major hot water usage centers located far apart, the latter is often a good idea anyway, from the standpoints of both convenience and economics. On the other hand, if you buy a tank that has far more capacity than you need, or has a much greater recovery rate than necessary, you will have plenty of hot water. However, you will be spending a great many extra dollars, both initially and in operating costs, for that convenience.

Instantaneous Water Heater

There is one other type of water heater that deserves mention here, one relatively new to this country but enjoying increased popularity in certain circumstances. Called an instantaneous water heater, it is both very small and a wall-mount unit that has no storage capacity whatsoever. It operates on either gas or electricity, and can be installed virtually anywhere. This heater requires only a cold water supply piped into one end, and a hot water pipeline can be attached to the other, or a faucet or shower head can be direct-connected. It operates as a flash boiler, instantaneously super-heating the water that passes through the unit. When a tap is turned on and the cold water commences to flow into the unit, it is immediately heated to a high temperature and passes through the outlet as steaming hot water.

The units themselves are fairly inexpensive and simple to install, and they operate very economically since there is no idling time or heat being dissipated and wasted from stored water. They are ideal for installation in summer camps or second homes, in recreational vehicles, or in new bathroom installations in houses, especially where the new bath location is remote from an existing hot water supply center. For that matter, several units could be installed individually at all hot water usage points in a new home. The initial expense would be moderately high for the units, but virtually the entire hot water distribution plumbing system would be eliminated, and operating costs drastically diminished.

The Distribution System

So then, the starting point of the residential hot water distribution system is at the water heater. Once the heater size and location are determined, fabrication of the piping system can go ahead. There are two types of pipe that can be used here. The one in most common use is CPVC, the same as for the cold water distribution system. The newcomer, PB, can also be used. The piping itself is done in the same manner as the cold water system, and in fact for the most part just follows right along with it. Be sure to keep the pipelines a minimum of 6 inches apart, both to allow yourself plenty of working room and also to prevent heat loss by transference from the hot line to the cold. Position the water heater so that is as close as possible to the major points of hot water usage, and run the pipelines in routes as direct as you can. If you have two or more locations of heavy hot water usage, calculate the benefits of installing two or more small tank-type or instantaneous water heaters and assembling two or more short hot water distribution subsystems.

The fittings and other odds and ends that you will need to complete the system are the same as for the cold water system. Additionally, you should install a pair of union fittings, one on the input and one on the output side, at the water heater. This will allow fast and easy disconnection should the unit need to be taked out for repair or replacement. Installing a pair of stop valves, one each on input and output, is also a good idea.

DRAINAGE DETAILS

If a water distribution system is poorly designed, incorrectly fabricated or sloppily installed, at least one can except some future headaches and aggravations from deficient operation and, at most, perhaps some property damage from leaks and such. But if a drain, waste and vent system is improperly installed, there will not only be future inconveniences and headaches, but also a problem of far greater proportion and consequence. An improperly arranged and installed DWV system presents a serious health and/or safety hazard that must be recokoned with. The job must be done correctly if it is to operate efficiently and effectively and present no hazards of a very real sort to the building occupants.

Fortunately, it is not one bit more difficult, and it costs no more in either effort or money to install a good DWV system rather than a poor one. There is absolutely no reason to skimp on materials, take shortcuts or make jury-rigged installations that are in violation of the plumbing code. Doing so is certainly not worth the potential dreary aftermath of insidious illness or other difficulties, particularly in view of the fact that such a system would have to be totally rebuilt anyway.

The basic rules of plumbing (and code) that affect the average residential DWV system design and installation will be touched upon. If your installation is a large one and/or has numerous complexities, you may run into some design and/or installation situations not covered here. In that case, by all means investigate the situation further until you are satisfied that all is in order. Matters pertaining to the design and installation of sewer lines and septic systems will be investigated in Chapters 7 and 8; our concern here is only with the DWV system that lies within the house itself.

DWV Purpose

We will begin with a couple of generalities, which include yet another cautionary note. The purpose of a drainage system is to exit human and nonhuman waste, both liquid and solid, from the house. In order to do so through a piping system, vents must be included in that system for proper flow. In order to accomplish this purpose, then, the first basic is to design and then install a system of pipes that will carry drainage from all the drain outlets within the house quickly and efficiently, with no leaks or hangups, out of the house and away. At the same time provision must be made to let air into the system as liquids drain out.

The system must be so constructed as to prevent backups of gases and raw sewage and to protect the occupants of the house from the harmful effects that could ensue. Raw sewage, of course, is laden with harmful bacteria of all sorts. It must be moved out of the house and away with no backups or leakage along the route. Gases are also almost always associated with sewage. You're familiar with the explosive properties of gasoline vapor, and doubtless know that methane is every bit as volatile. Hydrogen, of course, is extremely explosive and acetylene is the same stuff used in welders' torches. All of these gases and others as well are frequently found in sewer pipes. Alone or in various combinations they can be explosive, lethal and in the case of metallic piping (but not plastic), corrosive as well. So the second purpose of a DWV system is to keep this noxious stuff out. If you design and assemble your DWV system according to accepted practices and with the proper materials, both purposes of the system will be automatically satisfied.

DWV Pipe Type and Size

There are two choices in DWV plastic piping for use as drain-pipes, stacks and vents. The first is PVC, a piping that has excellent characteristics for this purpose. Also in common use is ABS, though this material is somewhat less satisfactory. Once you have settled upon which type you would prefer to use, the next step is to determine the various sizes of pipes that will be needed. Drainage pipes are sized according to their optimum drainage capacity in terms of a measure called a *fixture unit*, the equivalent of one cubic foot of liquid per minute of flow. Each plumbing fixture is assigned, as you can see in Table 6-5 a particular number of fixture units. These can then be related to the correct pipe size.

To begin with, always use the rule of thumb that an individual toilet should be connected to a 3-inch drainpipe, and a shower to a 2-inch drainpipe. Other individual items can be served by 1½-inch drainpipes. Some drainpipes will collect liquid from more than one fixture. Add up the fixture unit total for the line and find the proper size for either horizontal or vertical lines. In most cases, the soil stack must be sized at a minimum of 3-inch diameter in most locales, though some places require a 4-inch minimum. In lieu of local specifications, choose the 3-inch size, preferably with an outside diameter that will fit neatly within a conventional 2 × 4 stud wall space for ease of system fabrication.

The soil vent portion of the stack should be of the same size, though some codes will permit 2-inch diameter. But in cold weather climates, the portion of the soil vent (as well as any other) that protrudes above the roof should be at least 3-inch diameter and preferably 4-inch, to prevent frosting over.

Vent stacks that serve one or more plumbing fixtures are also sized according to fixture units. The smallest size, 1 ¼-inch, can only be used with an individual lavatory or an individual floor drain,

Table 6-5. Fixture units for calculating drainpipe sizes.

Fixture	Units
Kitchen sink	2
Dishwasher	2
Washing machine	2
Laundry tubs	2
Lavatory	1
Bathtub	2
Shower	2
Toilet	3
Bar sink	1
Darkroom sink	2

with no other fixtures attached to the vent line. The 1 ½-inch size is generally considered minimum for vent stacks, and will serve eight fixture units where sanitary tees are employed for connections, or 12 when wyes are installed. The 2-inch size will accommodate 16 and 36 fixture units respectively, and is the largest size likely to be encountered in most residential applications.

As to the main house drain, provided there are no more than two toilets involved, 3-inch pipe is satisfactory. If there are more than two toilets, the required pipe size is 4-inch.

Drainpipe Lengths

With the basic pipe sizes in hand, the next step is to take a look at drainpipe lengths. Soil and vent stacks present problems in residential applications only in the rarest cases, since a 1 ½-inch vent stack, for instance, would have to be more than 65 feet long total before exceeding code requirements. Horizontal drainpipes, however, are a different matter. You are limited in the length of drainpipes, depending upon size and pitch, not only by the plumbing code but also for practical reasons. If the total length and drop (in pitch) of the given pipe diameter is too great before it reaches a stack or is vented, the rushing liquid can build up such vacuum as it travels down the drainpipe that it will siphon the water out of the trap behind it and open the trap seal.

This critical pipe length is determined in terms of pipe diameters. You cannot connect a plumbing fixture any closer to a stack or vent than two pipe diameters, nor any farther away than 48 pipe diameters. For our purposes here, all you have to remember is these maximum lengths: 1 ¼-inch drainpipe, 5 feet; 1 ½-inch drainpipe, 6 feet; 2-inch drainpipe, 8 feet. The 3-inch diameter waste pipe that you will use to connect the toilet to a soil stack can have a maximum length of 12 feet, but it is always wise to keep that particular drainpipe as short as possible, down to a length of 6 inches but no less. None of your horizontal drainpipes should be any longer than the figures given, as measured from the trap outlet connection to the nearest wall or a vent or waste stack.

There is one circumstance where the maximum permissible length of drainpipes must be even shorter than this. This occurs when a series of plumbing fixtures, not including a toilet, are all connected to one branch drainpipe. The drainpipe runs to a soil or waste stack, and the fixture at the farthest end of the drainpipe line is vented through a separate vent stack of substantial size, which in turn provides venting for the remaining fixtures on the line (Fig. 6-13). This process is called wet venting, and though convenient to use in some plumbing layouts, it is not always permissible use. Check this before going ahead. If wet venting is permissible, the maximum drainpipe lengths are as follows: 1 ½-inch drainpipe, 2 ½ feet; 2-inch drainpipe, 3 ½ feet; 3-inch drainpipe, 5 feet. Again, the pipe length is measured from the connecting point of the trap outlet to the nearest wall of the serving waste or vent stack.

Now let's suppose that on your original plumbing layout you had some drainpipe lines that are longer than these specified maximums. What then? There are a couple of possibilities. One is to shift the location of the plumbing fixtures about in order to satisfactorily shorten the drainpipe lines. This is a good solution if you can arrange it, not only because it is likely to provide better drainpipe flow, but also because it will require less work in the installation and less cost for piping. Also, rearrangement in some cases means a possible reduction in the diameter of branch drainpipes, by virtue of splitting up the fixture-unit load on certain drainpipes or sections thereof. Another possibility is to introduce additional soil or waste stacks at appropriate points, or additional vent branches or stacks, thus effectively shortening the drainpipe lengths. This, of course, means additional work, piping and expense. But in many circumstances this is unavoidable.

Drainpipe Pitch

Now let's consider pitch. Obviously, all drainpipes in the entire system must be continuously pitched down hill and away from the building in order to effect proper drainage flow. All horizontal branches must be sloped slightly downward from fixture to soil or waste stack. Waste drops down the stacks into the main house

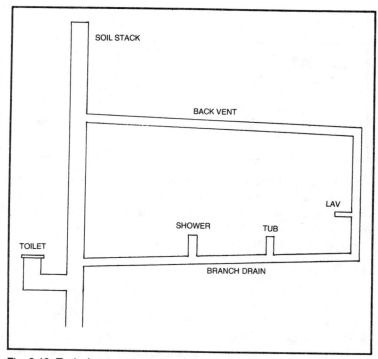

Fig. 6-13. Typical wet-venting arrangement.

drain, which also pitches downward into the sewer line. Various pitches are permissible and operate effectively, and the optimum degree of slope is dependent upon the drainpipe diameter. For our purposes in residential applications, the simplest approach is to remember that both the minimum and the maximum drop in inches of a horizontal drainpipe, measured from the trap outlet connection to the stack inlet connection is exactly the same as the diameter of the pipe being used for the three most common drainpipe sizes in residential use: 1¼-inch, 1 ½-inch and 2-inch. For the 3-inch size, the maximum drop is the same as the pipe diameter, 3 inches, and the minimum is just half that (Table 6-6).

A different, and perhaps easier, approach lies in using the same pitch everywhere in the drainage system. The permissable range of pitches run from ⅛-inch per running foot of pipeline to ½-inch, depending upon pipe size. However, if you simply stick with the optimum pitch, ¼-inch per running foot, you'll encounter no difficulties regardless of pipe size. Never vary this pitch, but keep it constant. If for some reason you cannot make up the pipe run with a

Table 6-6. Drainpipe lengths and drops.

Pipe Size	Drop (inches) Min.	Max.	Length Min. (inches)	Max. (feet)
1 ¼	1 ¼	1 ¼	2 ½	5
1 ½	1 ½	1 ½	3	6
2	2	2	4	8
3	1 ½	3	6	12
4	2	4	8	16

constant pitch because the necessary connection points do not line up, don't change the pitch to an inch per foot or something similar. Run the first part of the drainpipe at ¼-inch, and at the appropriate point drop into the stack with a 45-degree angle change by installing a couple of extra fittings (Fig. 6-14).

Venting Arrangements

Venting procedures seem to cause the most confusion, but in reality this can be done quite simply in nearly all residential applications. In cases where the system consists of one tightly grouped collection of plumbing fixtures, they all can be drained directly into a soil stack (Fig. 6-15). The grouping may be one bath, two baths back to back, a kitchen and bath back to back or whatever. The upper portion of the soil stack itself serves as the vent stack, provided that no other fixtures are connected higher on the stack than the group. This is called direct venting.

Another possibility is the wet venting mentioned earlier. In this case the toilet drains directly into the stack, and a common branch drain for the remaining fixtures enters the stack slightly above. The toilet is vented by the soil stack, while the fixtures are vented by another stack located at the far end of the branch drain line (Fig. 6-16).

A circumstance often arises where most of the plumbing fixtures are direct vented through the soil stack, but another lonely fixture is located at some distance away. If the distance is more than the permissible drainpipe length, individual venting may be the answer. This involves continuing the fixture drain line upward past the fixture and venting it out through the roof (Fig. 6-17). Thus, the fixture has its own individual small vent stack.

Another commonly used method in this situation, for one or several plumbing fixtures, is the back vent. In this case, the fixture drainage line continues up past the fixture for a minimum distance of inches above the drainage overflow level of the fixture. The pipeline then takes a turn at any convenient point and travels back to the main soil stack and connects to it (Fig. 6-18). There may be another plumbing fixture attached to the stack at a higher level than the one being back vented. If so, the vent pipe must rise beyond the lower fixture to a point of connection to the main stack at least 6 inches higher than the overflow drain level of the higher fixture (Fig. 6-19). Often several fixtures are vented through the main stack in this way, and this is called *loop venting*.

A residential DWV system usually is installed with one or another of these basic venting sytems. However, there is no reason why combinations cannot be used, or the same systems used a couple of times. For instance, a large, sprawling ranch house might have a kitchen-bath grouping of plumbing fixtures at one end, and another master bath-guest bath fixture grouping at the other. Here a direct venting system could be used for each grouping, each

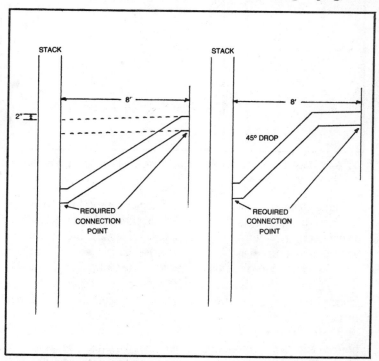

Fig. 6-14. If a drainpipe cannot be run at the proper pitch between required connection points, do not run it at a steep angle of less than 45 degrees (left). Instead, run the first part of the pipe at the proper pitch, then drop to the connection point at a 45 degree angle (right).

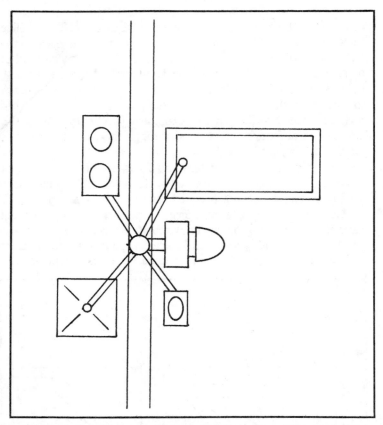

Fig. 6-15. Typical direct-vent piping arrangement where all fixtures drain into and are vented by the same soil stack.

having a main soil stack, with both stacks routed into a main soil stack, with both stacks routed into a main house drain (Fig. 6-20). A direct venting system could be used at one point in the house, and a back venting system used to vent a scattering of additional plumbing fixtures to another main soil stack at a different point (Fig. 6-21).

There is yet another possibility for venting that is helpful where there are several groupings or scatterings of plumbing fixtures, and this involves the installation of a vent stack. The purpose of a vent stack is only for venting, not for carrying wastes. Its bottom end may be attacked to a horizontal branch drainpipe or to a stack. The upper end passes through the roof to the outdoors. To this vent stack any number of branch vent lines can be attached to serve any number of plumbing fixtures in a back venting or revent-

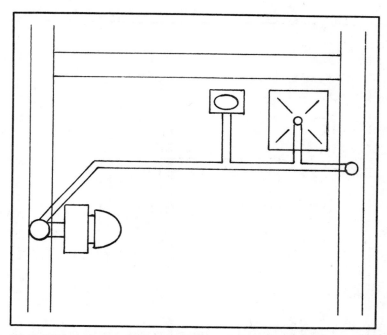

Fig. 6-16. Typical wet-venting arrangement with toilet venting and draining to soil stack, shower and lavatory draining to stack and revented by separate vent.

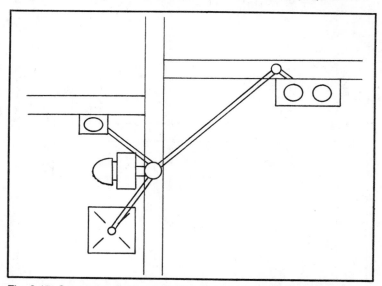

Fig. 6-17. Combined venting methods with fixture group vents and drains to stack; remote fixture has individual vent stack.

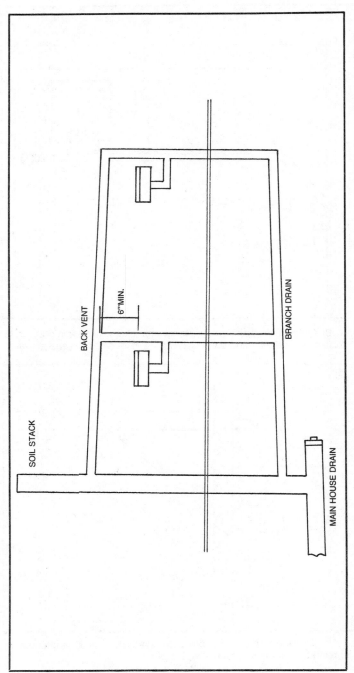

SOIL STACK

BACK VENT

6"MIN.

BRANCH DRAIN

MAIN HOUSE DRAIN

Fig. 6-18. If there are no higher fixtures, a back-vent connection must be made no lower than 6 inches above the flood or overflow level of the fixture nearest the vent stack.

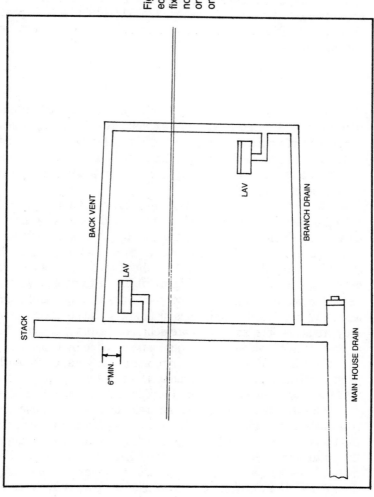

Fig. 6-19. If higher fixtures are connected to the stack, the back vent for lower fixtures must be connected to the stack no lower than 6 inches above the flood or overflow level of the highest fixture on the stack.

249

ing arrangement (Fig. 6-22), provided only that the size is of the vent stack is sufficient to handle the number of attached fixtures. In residential applications, 1 ½-inch diameter pipe is generally sufficient, although 2-inch pipe is sometimes used.

Here's a further point about vent pipes. While drainage pipes slope down and into the stacks, horizontal branch vent pipes slope upward and into the stack, using the same degree of pitch as for the drainpipes, or ¼-inch to the running foot. This is because the inflow of air comes down the stack from the top and into the drainage lines. Also, the upsloping vent lines allow gases a natural upward escape path, and they cannot become entrapped. The lengths of these vent lines can be whatever is practical, but they should always be kept as short as possible in the interest of an efficient system and an economical installation.

Traps

The final major element of the DWV system that needs attention involves the traps. With the exception of the toilet, which contains an integral trap, every single drainage fitting on a plumbing fixture must be fitted with an appropriate trap. The most common ones in use are variously called S, half-S, P or U depending upon their general shape.

Which one you use is a matter of the physical characteristics of the particular installation. A full S trap would be used where a kitchen sink drains down through the floor (Fig. 6-23), for instance. If the sink drains into the wall above floor level, a P trap might be used (Fig. 6-24). Another type of trap, the drum trap, is sometimes used with shower installations (Fig. 6-25). These traps are always a second choice and used only where other types are impractical from a physical standpoint. Drum traps must always be installed where they are accessible, since they are debris collectors and cannot be cleaned out with an auger. The job must be done by hand.

You will need one trap of appropriate configuration for each plumbing fixture, together with suitable lengths of tailpipe to connect the trap to the drain outlet. There is one exception to this. A double-bowl kitchen sink that has two drain outlets can be connected to a single trap, using a fixture drainpipe assembly that is made for this purpose. Solid, one-piece traps should contain a cleanout plug at the bottom of the bend. Sectional traps that come apart in two or more pieces by removing the slip nuts, however, can be easily disassembled for cleaning and need no plug. Traps are sized accord-

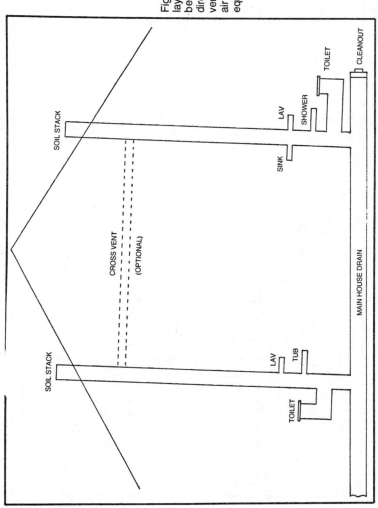

Fig. 6-20. In a sprawling plumbing layout, two separate soil stacks may be required, shown here with direct-vent fixture groupings. Cross vent is sometimes installed for better air and drainage flow and pressure equalization.

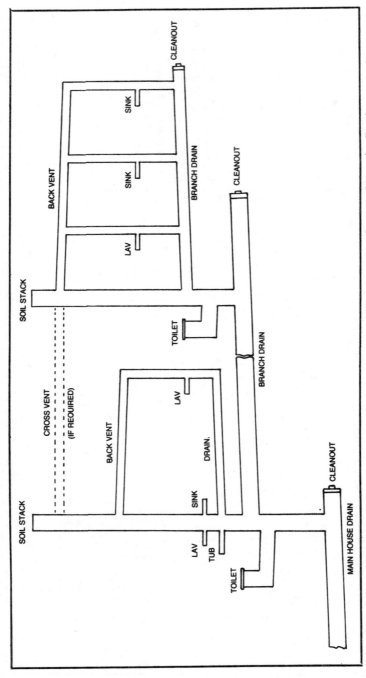

Fig. 6-21. Complex DWV system with two soil stacks, back venting, loop or circuit venting, direct venting and optional cross vent.

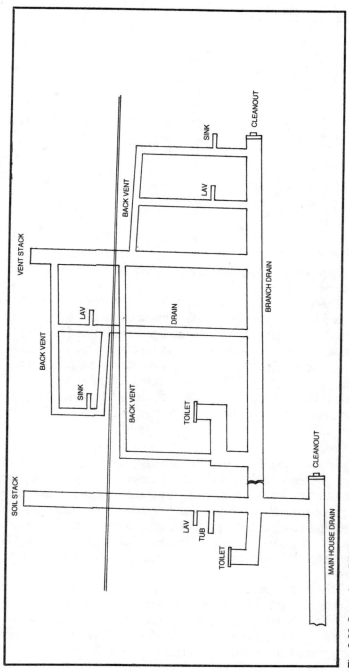

Fig. 6-22. Spread-out DWV system may require soil stack, extensive branch drainpiping, plus a separate vent stack to handle back vents and circuit vents.

253

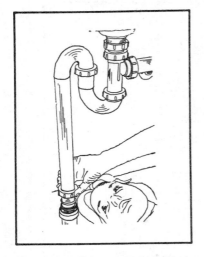

Fig. 6-23. S-trap is used to connect fixture to drainpipe stubbed up through floor.

ing to pipe size; a 1 ½-inch kitchen sink drain is fitted with a ½-inch trap, and so on.

There is one other point in a residential DWV system where a trap of a somewhat different nature might be used. This is a large U-trap having a cleanout plug at the top of each arm of the U. It is generally located in the main house drain close to the point of exit from the house (Fig. 6-26). It serves as a master water seal, so to speak, for the entire DWV system that lies behind it within the building. It also captures a certain amount of solid material, and may have to be cleaned out from time to time.

DWV Layout

When laying out your DWV system, there are a number of points to keep in mind. As we've mentioned before, keep the pipe runs short and simple and maintain the constant drainage pitch. The drainpipe runs should contain as few fittings and make as few directional changes as possible. Drainpipes from fixture to stack should turn no more than 90 degrees total, if you can possibly manage it. This takes plenty of forethought, but usually can be done.

Throughout the entire system, arrange everything for the best possible gravity flow by providing as gentle a path as you can for the liquid. For instance, if you have to make a right-angle turn, don't use one abrupt 90-degree elbow to do the job. Instead, install two 45-degree fittings with a length of pipe between them, or use a long sweep elbow (Fig. 6-27).

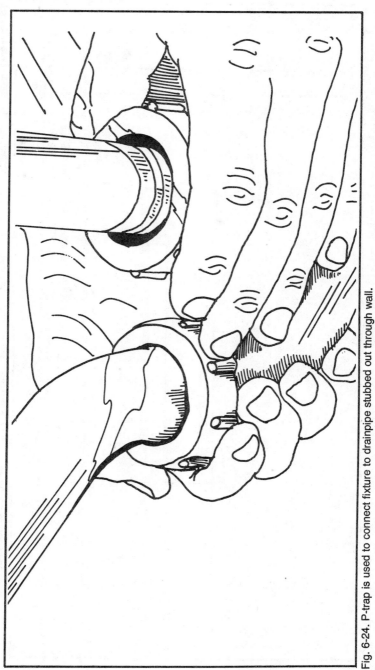

Fig. 6-24. P-trap is used to connect fixture to drainpipe stubbed out through wall.

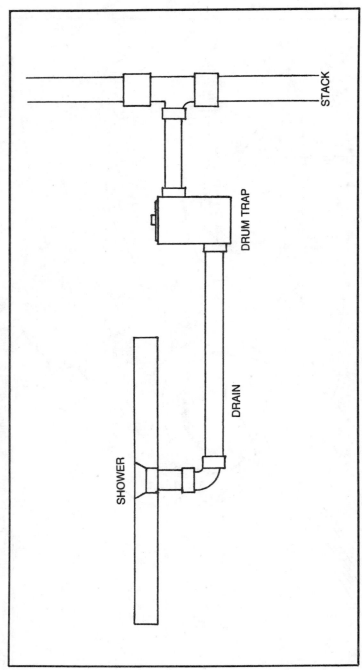

Fig. 6-25. Typical drum trap arrangement for shower installation. It may also be used for bathtub or other fixtures, but generally is not.

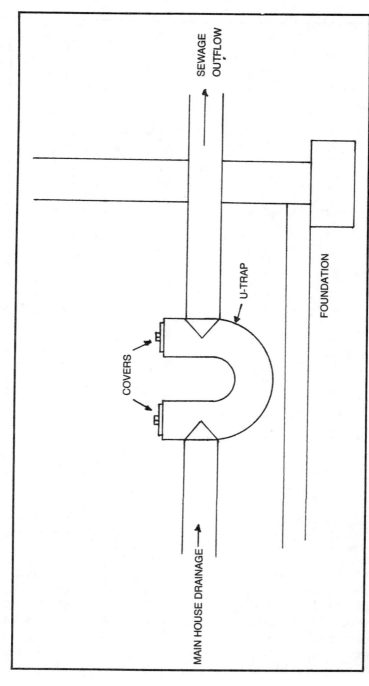

Fig. 6-26. Large U-trap is sometimes installed close to the point of exit of the main house drain through the foundation.

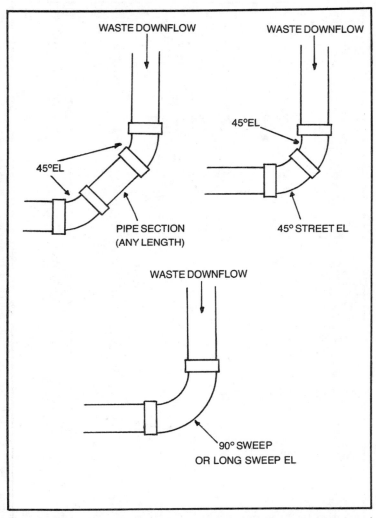

Fig. 6-27. Three methods of piping to lessen abruptness of direction change and improve flow characteristics in vertical-to-horizontal DWV line.

When the waste flow makes a change from a horizontal drain-pipe to a vertical stack, or simply a vertical drop, an abrupt change is all right. Thus, a 90-degree elbow could be used, or a sanitary tee (Fig. 6-28). A common tee should not be used, since it does not have the downward-curved corner that provides good flow. If the flow is from vertical to horizontal, down a stack and into a drain line, an abrupt transition should never be used. Here, a 45-degree elbow

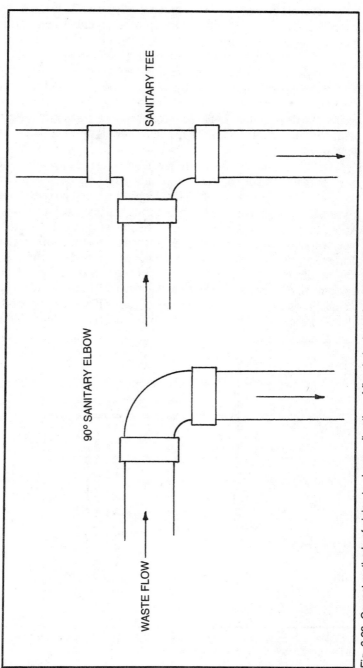

SANITARY TEE

90° SANITARY ELBOW

WASTE FLOW →

Fig. 6-28. Correct methods of piping to change direction of flow in horizontal-to-vertical DWV line.

in the vertical line makes the transition gently into a Y-fitting at the end of the horizontal line (Fig. 6-29). Since an abrupt change of this sort should never be made without the installation of a cleanout plug, the other arm of the Y provides that opportunity.

Always consider the direction of the waste flow and make the flow path as easy as possible by using properly curved fittings. Watch out for mistakes such as installing a sanitary tee upside down, against the flow. But note that for vent purposes, a sanitary tee can be used upside down, since the (air) flow here is upward instead of downward.

Laying out your DWV system can be done in much the same manner as the cold or hot water distribution system. Simply choose the best and most direct routes for the pipeline, and go over them section by section, noting what fittings you will need along the way. At the same time, jot down the approximate lengths of the various pipe sections, so that you will know how many stock lengths of what diameters you must purchase. You will doubtless find that there are two or three different ways to arrange certain segments of the system, all of them apparently satisfactory as far as operation and installation is concerned. In that case, choose whichever happens to appeal the most.

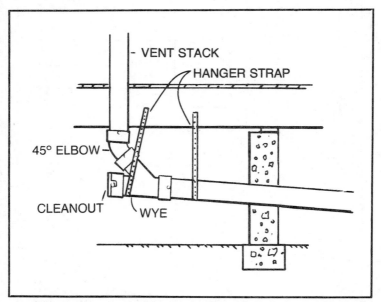

Fig. 6-29. This common method of transition from vertical drainpipe to horizontal line provides the necessary cleanout (courtesy of Genova, Inc.).

If you run into a problematical section, don't be afraid to ask your supplier a few questions. He probably can provide you with some sound, workable answers. As you lay out your system, you may also find it useful to work up some rough diagrams, either in floor plan, in elevation or both, that will show you the basics (though not the specific pipe runs and installation details) of the system. This often makes visualization of the entire system a bit easier, and helps to eliminate some of the potential for mistakes.

Chapter 7

Septic Systems

If you are in a position to be able to tie into a municipal sewage system, this chapter may not be of much interest to you and you can skip ahead. But anyone who opts for rural living, or for that matter in any area where a municipal sewage system is unavailable, must provide for his own sewage disposal system. Though some of the so-called waterless, composting or biodegradable waste disposal systems for residential applications are now making an appearance in this country, they have yet to gain widespread acceptance with either the general public or with building departments or building code writers. In time this situation is likely to change considerably. For the moment, though, the most acceptable and practical arrangement for individual sewage disposal, unsatisfactory though it may be, is the septic system.

SEPTIC SYSTEM DESIGN

The design of a septic system for a residence is undertaken in somewhat different fashion than the other parts of the plumbing system found within the building. Before anything can be done a certain amount of preliminary information must be gathered up from several sources. A determination must be made as to whether or not a septic system can actually be installed on the property and made to work effectively. By rights, this task should be accomplished in the earliest stages of planning for a new home, perhaps even before the property has been bought.

Preliminary Investigation

In essence, a septic system operates by converting all of the waste water and solids from the house into liquid, gases and a small residue of solids. The gases, which account for only a very small portion of the total volume of sewage, escape into the atmosphere. The solid residue, or *sludge*, remains in the tank and builds up very slowly. It periodically must be removed and disposed of off the premises. The remaining liquid, or *effluent*, which constitutes by far the largest portion of the volume of material in the tank, is slowly released into drainpipes, on a more or less continuous basis. The *effluent* disappears by absorption into the soil and evaporation into the air. Most of the liquid, and in a few designs essentially all, disappears into the ground. The ground, then, must be capable of fully absorbing all of this liquid and draining it safely away. Otherwise, the result would be a fetid and noxious swamp.

Before you purchase a piece of property with an eye toward building yourself a new home, make sure that the absorptive capacity of the soil is sufficient to handle the effluent from a septic system. At this point no septic system design is necessary. All you need to know is whether or not you will be able to install one of reasonable capacity for residential use. There are several ways you can go about getting this assurance.

If there are already other houses in the area, you can assume that they, too, are operating with septic systems, probably with a workable degree of success. Talking with one or more of the homeowners would be a good idea, and would give you a direct reading of any problems that might have been encountered in the existing septic system installations or in their current operation.

Another possibility is to go to the local building department offices and inquire as to what the situation might be on or around the particular piece of property you are looking at. If there are other homes in the immediate vicinity, those officials should be aware of any problems that might exist. Local or state health departments or similar offices might also be able to give you some information.

If you plan to build in a remote area and there are no existing homes nearby, making a solid determination is more difficult. You might be able to locate a soil map of the area in question. These maps give complete descriptions of the soil and its characteristics. Note the different kinds of soil and outline some information about soil qualities and potential uses. The maps (which are not available for all areas) can be obtained from the Soil Conservation Service in Washington, D.C., or through their local offices. State or local agriculture experimentation stations or similar agencies can also be contacted.

You can also request a local building inspector or health department inspector to look over the premises and give you an opinion as to whether or not a septic system can be installed. Failing that, you can hire a soil engineer—the Yellow Pages list them— to make a soil survey. He will give you a written analysis of his findings as they relate to soil absorption and the potential efficacy of a septic system installation in that particular location.

In all cases, the question you have to ask is a very simple one. "Can I install a residential septic system here, and if so, can I expect it to work effectively?"

If, after making your investigations, you feel that there is still some uncertainty about the whole affair, there are two more steps that you can take. One is to insist that percolation tests be made upon the property, either by yourself or by others, to prove adequate soil absorption for a septic system installation. Obviously, if the tests are negative you'll not want the property.

The second item is to demand that the property purchase contract that you are about to sign contains contingency clause. This clause should state that if the soil absorption rate is inadequate for a septic system installation, the contract is thereby invalidated and you don't have to carry through with your purchase.

Preliminary Design Parameters

Once you have made the determination that a septic system installation can indeed be made upon your property, the next step is to ascertain what your preliminary design parameters must be. Most of this information is predetermined, as a rule, by the building department or other governing bodies, such as the State Health Department, that has jurisdiction in your area. They will have certain requirements that you must follow, so you'll have to go to them to get the necessary information.

They will doubtless want to know such things as the size of the family that will occupy the residence, the number of bedrooms in the house, perhaps the number of fixtures that will be attached to the plumbing system and drain into the septic system, or similar details. They will tell you what minimum septic tank size you must use, the minimum and perhaps the maximum distance the tank must be located from the house, the distance the tank and/or absorption field must be located away from property lines and water sources, and items of that sort. There may also be requirements as to septic tank material, system layout or dimensions, piping methods or materials, etc.

In the absence of official instructions, specifications, or suggestions, you may have to work out some of these initial details yourself. You'll need four basic elements: the *house sewer line* that attaches to the main house drain, the *septic tank*, the *sewage outflow line*, and the *absorption* or *leaching* pipelines. The tank is frequently placed about 10 feet from the building wall, sometimes a bit less and sometimes more. The sewage outflow line that travels out of the opposite side of the tank can be of almost any practical length, but generally is kept as short as possible. At the end of this line will be the absorption field, which can go straight on out or be cocked off to one angle or another. At this point, you have no idea how large that field must be. You can be assured, though, that it will take up a considerable amount of space.

Now you need to formulate some rough idea as to where everything is going to go. To begin with, you'll need to determine about where the house will be situated, and where the main house drain will emerge if those plans have already been made up. This will determine approximately where the septic tank must be located. Then you have to make sure that there is room for a large absorption field.

If there are obstacles in the way that seem insurmountable, you may have to move the house site, shift it about, or perhaps change the exit location of the main house drain. This will in turn necessitate some changes in the plumbing plans. If those plans have not yet been drawn, you have a freer hand. Remember that in the interest of economy the main house drain should exit the house as close as possible to the soil stack, which in turn should be as close as possible to the central areas of water usage and waste disposal within the building.

There are a few points to keep in mind as you try to spot the locations for all of this equipment. First, take note of any setback requirements that restrain you from going too close to a lot line or easement. Remember that the septic system should be as far as possible from a well, and at least 100 feet away and on the downhill side of any water source. Consider that the sewer line and the sewage outflow line will require trenches, and the septic tank will rest in a large hole. This means you will need room to dig and room to stack the spoil. You will want to avoid huge boulders, nice old trees and such wherever possible. The leaching field installation will require tearing up a great deal of ground, and you'll need maneuvering room here, too.

Leaching Field

A leaching field can be built on flat ground or on slightly sloping ground with no trouble. But a slope in excess of 15 degrees makes things much more difficult, and is to be avoided if possible. Also, the shallow type of leaching field should be planted to grass or hay, and should not be heavily shaded by thick shrubbery or tall trees. You need open ground to work with, and it should remain that way when you are done. Bear in mind, too, that the leaching field should be free from rocks and boulders that will make the installation expensive and difficult. If there is underlying rock strata less than 4 feet beneath the ground surface, the absorption field may not work well, if at all. In addition, if your prospective leaching field area has a high water table, or is apt to be saturated with water at certain times of the year, forget it. The site won't work.

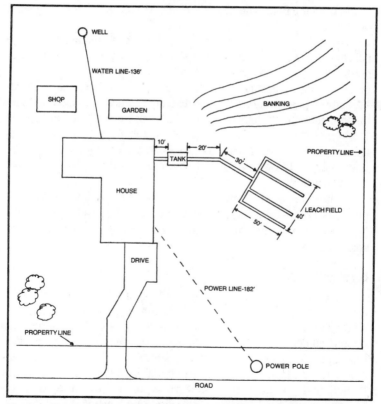

Fig. 7-1. Plot plan type of layout drawing showing principal elements of water supply line and waste disposal system and their locations.

Once you have all of these preliminary facts and details pulled together, along with any others that might pertain or have some effect upon the system, you can work out a rough block diagram that shows approximately what you need to have (Fig. 7-1). This diagram should show the approximate relationship of the septic system elements with the house and any other nearby improvements, as well as natural terrain features. With this accomplished, you are ready to begin on the next step.

PERCOLATION TESTS

Percolation tests are used for two principal reasons. The tests determine the general absorption capabilities of the soil in the septic system area and the specific absorption capabilities of the soil where the leaching field is to be built. The latter rating determines the size of the leaching field, and in some cases may determine the location or the directional lie of the field as well.

Percolation tests may or may not have been made prior to the purchase of the property or before construction work began on the house. The tests may or may not have been made at the actual site of the proposed leaching field. Even if tests have been made sometime in the past, local authorities usually require that such tests be made again shortly before the field is constructed. In some cases the homeowner can make the tests himself. It is a simple enough job. In others cases it can only be done under supervision of the governing body or perhaps only by authorized employees of that governing body.

There are a number of ways in which percolation tests can be made, but they are all variations on a theme. They all serve to test the same thing—the rate at which water can be absorbed by the soil. Here are the basics of how the task is performed.

Digging the Holes

The first step is to pick any handy spot within the confines of the proposed leaching field area. Dig a hole down to the depth at which the drainage pipes will lie. This depth is variable depending upon local conditions and regulations. Most fields are kept as shallow as possible to aid in evaporation, sometimes only 12 to 18 inches below the surface. But the pipes should lie below the level of the topsoil, and in some areas of severe winter weather they must be placed lower than 18 inches. Be guided by local custom.

Whatever the case, dig the hole big enough so you will have room to work, and flatten the bottom of the pit out. In the center of this pit dig another hole 12 inches square and 12 inches deep, square-sided and flat-bottomed. Roughen the sides of the hole with a flat piece of rock or some other implement to remove any tightly compressed earth slick that might have been formed by the shovel. Place an inch or two of sand or fine gravel in the bottom of the hole.

Getting a Water Supply

The next step requires that there be an adequate supply of water on hand. If you can arrange to have a garden hose handy, that's best. If not, you'll have to haul in trash cans or drums full of water. Fill the 1-cubic foot hole almost full of water, taking care not to crumble the top edges inward as you do so. At the end of half an hour take a measurement to see how far the water level has dropped. After making the measurement, top the hole up again and repeat the procedure. This should be done regularly for at least 4 hours, with the water level drop noted each time. Add the total number of inches together and divide by 4 to get the average rate of drop per hour. The resulting figure is the percolation rate.

Tests and Test Hole Arrangement

There are plenty of alternate methods of making such tests. For instance, the hole may simply be kept topped up, or at least relatively full, for a period of about 4 hours, with the total amount of drop noted after another 2 hours have passed. Or, the hole may be periodically topped up and checked over a 24-hour period, with level-drop measurements taken at a spaced intervals.

The test hole arrangement may also be somewhat different. One method is to use round holes of 4-6 inches in diameter, with a series of 6-12 holes scattered about the proposed leaching field site. Measurements are taken at each hole and combined to find an average percolation rate for the entire field. This is particularly advantageous where the soil consistency changes markedly from point to point. A series of test holes may be dug in a number of widely scattered spots to begin with, in order to find an area that has the best percolation rate. Then another series of test holes can be set up in an attempt to define the high-rate area so that the leaching field can be placed directly upon it.

Interpreting Test Results

Once the tests have been deemed satisfactory, the results can be interpreted and put to work. If the percolation rate is found to be less than 1 inch per hour, the chances are excellent that a sewage disposal installation will be disallowed. The soil simply is not capable of absorbing a substantial effluent discharge with any effectiveness, and the result would be a health hazard. A rate lying between 1 and 2 inches per hour is marginal, but workable, if the field is a relatively large one an properly and carefully installed. Even so, the field would probably have to be replaced or relocated after a few years of service. A rate range of from 2 inches to 4 inches per hour is much more acceptable, and a leaching field operating under these conditions should work well and have a good life span.

As the percolation rate increases to 6 inches per hour, or an inch every 10 minutes, the situation becomes more and more ideal. Percolation rates much higher than that are somewhat unusual, though they do occur. If you have a rate of an inch every 5 minutes, you can expect a leaching field of even small size to drain faultlessly for an indefinite period. A perc rate of 1 inch per minute is phenomenal but does happen.

Governing bodies that have control over septic system installations will place their own interpretations upon the percolation test results, and these may or may not equate with the figures just discussed. In some locales a 3-inch rate might not be sufficient in the eyes of officialdom, while in others a 1-inch rate might well pass.

The rate may also be used to determine the area required for the leaching field. Here, too, the way the figures are used may be arbitrary and certainly will be different from place to place. However, the requirement for a given locale might be based upon 200 square feet of leaching field area for every bedroom at a 6-inch rate, 250 square feet at a 4-inch rate, 300 square feet at a 2-inch rate, and so on. There may be some other basis for making the determination.

If there are no particular requirements, you'll need some sort of general guideline to work with. Generally speaking, if your perc rate is 6 inches or more per hour, 150 square feet of leaching field area for every occupant of the house should be ample. Use about 175 square feet at a 4-inch rate, 200 square feet at a 3-inch rate, 250 square feet at a 2-inch rate, and 325 square feet at a 1-inch rate. But in all cases no harm is done by making the field half again as large as the numbers might indicate. The expense for doing so at the time is slight, and the larger field is less likely to become befouled as rapidly as one of marginal size.

SETTING THE SPECS

Now you have a rough layout of the major components of the septic system, and you have complied most of the preliminary data.

With this information at hand, you can begin setting the specifications for the complete system, outlining in some detail the material you will use.

SEWER LINE PIPE

First you'll need a run of pipe of suitable type, diameter and length to serve as the house sewer line, running from the house drain to the septic tank. You've already determined the length on your layout plan. In plastic systems, a common procedure is to install a 3-inch diameter main house drain, but to make the sewer line of 4-inch diameter. A 4-inch by 3-inch adapter will join the two lines. If a cleanout has not already been installed in the main house drain inside the building, this is a good place to put in an above-ground cleanout plug, especially if the septic tank is far from the building. Several kinds of pipes can be used, with PVC or ABS plastic being common nowadays.

SEPTIC TANKS

Next comes the septic tank. In theory, a septic tank can be sized according to the maximum amount of waste produced by the users of the system in a given 24-hour period. Using a reasonable figure of 50 gallons of waste per day per person, a family of four could therefore get by with a tank capacity of 200 to 250 gallons, and there would be satisfactory septic action. However, in practice this rule of thumb is almost never followed. The U.S. Public Health Service recommends a minimum tank capacity of 750 gallons for a two-bedroom home. They consider a 900-gallon tank suitable for a three-bedroom home, and a 1000-gallon tank for a four-bedroom home.

Many local codes disallow the installation of anything smaller than a 1000-gallon tank in single-family residential service. Often the requirements are greater. In view of the fact that small homes are frequently converted to larger ones somewhere along the line, the minimum 1000-gallon size is a good idea in the absence of local regulations to the contrary. If you are building a particularly large house that may well have more than five occupants for length periods of time, adjust the tank capacity upwards to suit.

Composition of Septic Tanks

Septic tanks are available in a wide variety of configurations, and the best bet is to check what is available locally. They can be made of any one of several materials. You may find that one or two specific materials are either required or disallowed in your area. The least expensive and shortest-lived tanks are made of heavy asphaltum-covered steel in cylindrical shape, with a matching steel cover. Precast concrete tanks are becoming increasingly popular and serve well. These tanks may be rectangular or cylindrical and have reinforced concrete covers fitted with one or two clean out and inspection hatches. Fiber glass tanks are also available in various configurations.

Septic tanks can also be built up on the site. They might be constructed of precast concrete rings set on a poured concrete base, for instance, and capped with a precast concrete cover. Large rings of vitrified clay tile might be used, though these are usually confined to smaller tanks. Tanks can also be made on the job site from poured concrete. This involves building the necessary forms and then pouring the concrete from ready-mix trucks, after which the concrete is allowed to cure and the forms are stripped. This type of tank should not be put into service for a couple of months after construction, in order to allow complete curing of the concrete. Septic tanks are also sometimes made from standard concrete blocks, laid upon a poured concrete foundation and capped with a precast reinforced concrete cover.

If you decide to build your own septic tank as many people do, be sure to start you planning with a complete list of requirements and specifications for septic tanks from whatever authority has jurisdiction over such installations in your area. If there are no particular requirements, you may have to work up a set of plans (usually simple sketches including all pertinent construction details are sufficient) to submit for approval before beginning. Even if this is not necessary, you would do well to research the situation a bit. Rather than starting your plans from scratch, follow an existing set of plans that are known to be workable. Make minor modifications to them as necessary.

Siphon-Septic Tanks

Many septic tanks consist of a large single container of appropriate capacity, with an inlet at one side and an outlet at the other (Fig. 7-2). However, there is another type of septic tank of somewhat greater effectiveness, called a *siphon-septic* tank (Fig. 7-3). This arrangement has two main sections that may either be housed in one two-compartment unit or built as two separate but interconnected tanks. The first section is the septic tank itself, in this case

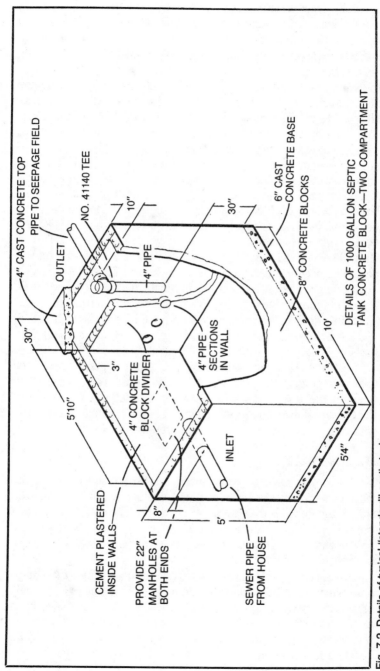

CEMENT PLASTERED INSIDE WALLS

PROVIDE 22" MANHOLES AT BOTH ENDS

SEWER PIPE FROM HOUSE

4" CAST CONCRETE TOP

NO. 41140 TEE

OUTLET

4" PIPE

PIPE TO SEEPAGE FIELD

10"

30"

6" CAST CONCRETE BASE

8" CONCRETE BLOCKS

10'

5'4"

DETAILS OF 1000 GALLON SEPTIC TANK CONCRETE BLOCK—TWO COMPARTMENT

30"

3"

4" CONCRETE BLOCK DIVIDER

4" PIPE SECTIONS IN WALL

5'10"

INLET

8"

5'

Fig. 7-2. Details of typical "standard" septic tank.

273

generally called a *settling* tank, and its construction and purpose is nearly identical to a single-unit standard septic tank. The second section is the *siphon* tank. In the compartmented single-unit design there is an integral submerged outlet which leads from the settling tank to the siphon tank. Where the two units are built separately, they must be plumbed together with a short pipeline. The siphon tank contains a trap at the bottom, with an overflow drainpipe attached to the outflow side of the trap. The inlet side of the trap is covered with a screened hood.

In operation, the settling tank works just the same as an ordinary septic tank. As waste is dumped into the tank, an equal amount of effluent flows into the siphon tank. When the level in the siphon tank reaches a certain height, the resulting pressure forces a large discharge of effluent through the trap and out into the leaching field until the tank is nearly exhausted by siphonage. Unlike the septic tank, the siphon tank empties itself periodically and all in one shot. The effluent discharge is under sufficient pressure that all parts of a properly designed leaching system will receive a largely equal effluent flow. After each discharge the field has a considerable period of time to absorb the discharge and recover for the next one. That time span is dependent upon both the size of the siphon tank and the amount of waste that is discharged from the house on a daily basis.

After exploring the various types of septic tanks and their pros and cons, you can then make an informed decision as to which type might by the best one for your particular installation. Make a note of the general specifications, including size and any details having to do with tank emplacement, of the tank you have chosen.

LEACHING FIELD

Now you have to jump ahead a little bit and consider the leaching field. You already have the figures needed to determine the size of the field, so that can be worked out. The square footage of the field is not equal to the entire area that the whole field will take up, but rather the area that the leaching trenches in which the pipes are placed will cover.

In a leaching field, technically known as a *subsurface-tile absorption system*, each individual run of drainpipe (sometimes there is only one) is contained in a separate trench. Trench width varies from locale to locale, or even from job to job, and in your area a certain trench width may be specified. However, 18 inches should be considered a minimum, and 24 inches is common enough. These

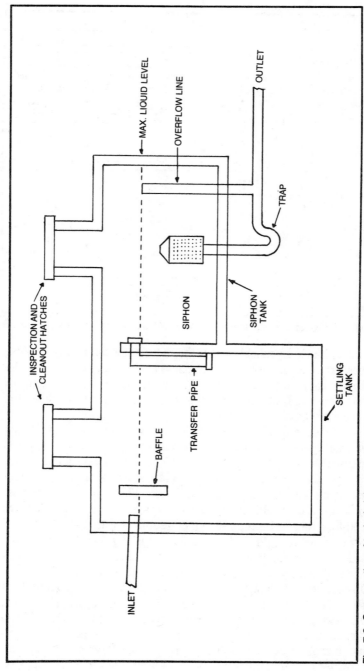

Fig. 7-3. Cross section of typical siphon-septic tank.

INLET

BAFFLE

TRANSFER PIPE

INSPECTION AND
CLEANOUT HATCHES

SIPHON

SETTLING
TANK

SIPHON
TANK

MAX. LIQUID LEVEL

OVERFLOW LINE

TRAP

OUTLET

275

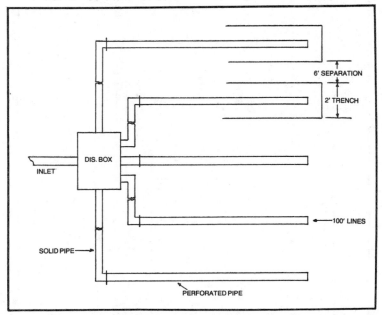

Fig. 7-4. Layout of typical leach field with 1000 square feet of leaching area.

two widths are probably about the most practical, if only because they are standard widths for backhoe buckets and that makes the digging easy. In order to determine the square footage of your leach field, you need only to choose a trench width and then work out the necessary trench lengths.

For example, if you need 1000 square feet of leaching field, that would require a trench 2 feet wide and 500 feet long. This length is impractical, though, and is also ineffective for a leaching system. Since individual drainage lines should not be more than 100 feet long, you will have to rearrange things. There are plenty of possibilities. You could simply divide the system up and install five 100-foot lines or better yet, ten 50-feet lines (Fig. 7-4). You could set up a square grid like the one shown in Fig. 7-5. This utilizes a solid, nondraining header, to which are attached nine outflow lines each 50 feet long. These are connected across the far side with a 48-foot run of drainpipe. The total square footage works out to just 4 square feet less than the required 1000. Or you could split the field into two sections, with half the leaching field running off in one direction and half in another. You could also use some other combination that might be feasible insofar as physical layout is concerned, provided only that the total amount of drainage trench area approximately equals 1000 square feet. If you end up with a little extra, that's no problem, but don't skimp.

Piping the Field

There are several different kinds of plastic pipe that you might use in piping the leaching field, and a good many brand names. For instance, ABS is popular for this purpose, and so are PE, PVC and SR. Solid pipe is used at points where drainage is not required or desired, while perforated pipe is used in the absorption sections of the field. Fittings used in a leaching field installation are the standard fittings designed for an compatible with your particular kind of pipe. Most of the pipe used is rigid, but there are some new varieties of flexible and semiflexible pipe, some smooth-surfaced and others corrugated. By and large, whatever your local lumberyard or plumbing supply dealer can handily provide you with will probably be suitable. If there are certain code requirements for such pipe that locally pertain, that's the kind of pipe he'll stock.

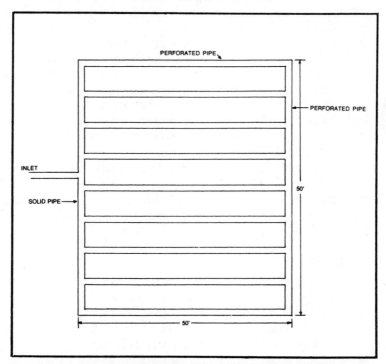

Fig. 7-5. Layout of grid-type leach field with nearly 1000 square feet of leaching area.

Seepage Pit and Deep Bed

There are methods of building absorption fields other than the subsurface-tile sytem. One is called a *seepage pit* or *seepage bed* (Fig. 7-6). This arrangement differs in several respects from an ordinary leaching field, in that very little reliance is placed upon evaporation. The drainpipes are buried deeper. They do not lie in trenches but rather in a single extensive bed of gravel. Seepage beds to do have some advantage. They are easier to build in some instances, are very effective under proper soil conditions, and the pipes lie well below the surface where they are less suspectible to mechanical damage or freezeup problems. However, seepage pits are prohibited in many areas, so check before you begin drawing up plans.

Another possibility is an installation called a *deep bed*, or *deep seepage bed* (Fig. 7-7). The construction elements of this type of field are quite similar to that of the seepage pit, but the bed is set far deeper into the ground, often as much as 15 feet or so. This arrangement might be used, for instance, where the topsoil or subsoil layer, or both, runs exceptionally deep and does not have good percolation capabilities. By digging down to a lower level, the field can be placed upon a stratum of gravel, up-ended broken shale or some similar geologic layer that has good percolation capabilities. The presence of such layers can be determined, along with their depth and thickness, by taking core-drilling samples. If a high-perc layer does exist within a resonable distance below the surface of the ground often it is well worthwhile to install a deep bed. Obviously the smaller the field can be the better.

As far as the plumbing is concerned, a seepage pit or a deep bed is made up in much the same manner as an ordinary subsurface-tile system. Plastic pipe in 4-inch diameter is used along with the appropriate fittings. The drainage lines themselves are perforated, while connection sections are solid.

SEWER OUTLET

With the leaching field details out of the way, you can now jump back a step and turn your attention to the *sewer outlet* or *effluent outflow line*, the one that joins the outlet of the septic tank to the leaching field drainpipe. The pipe used here is of the same kind and diameter as that used in the leaching field, of the solid variety. You'll need sufficient pipe and fittings to go directly from the septic tank outlet to the field.

If the field consists only of a single-drainpipe line, all that is

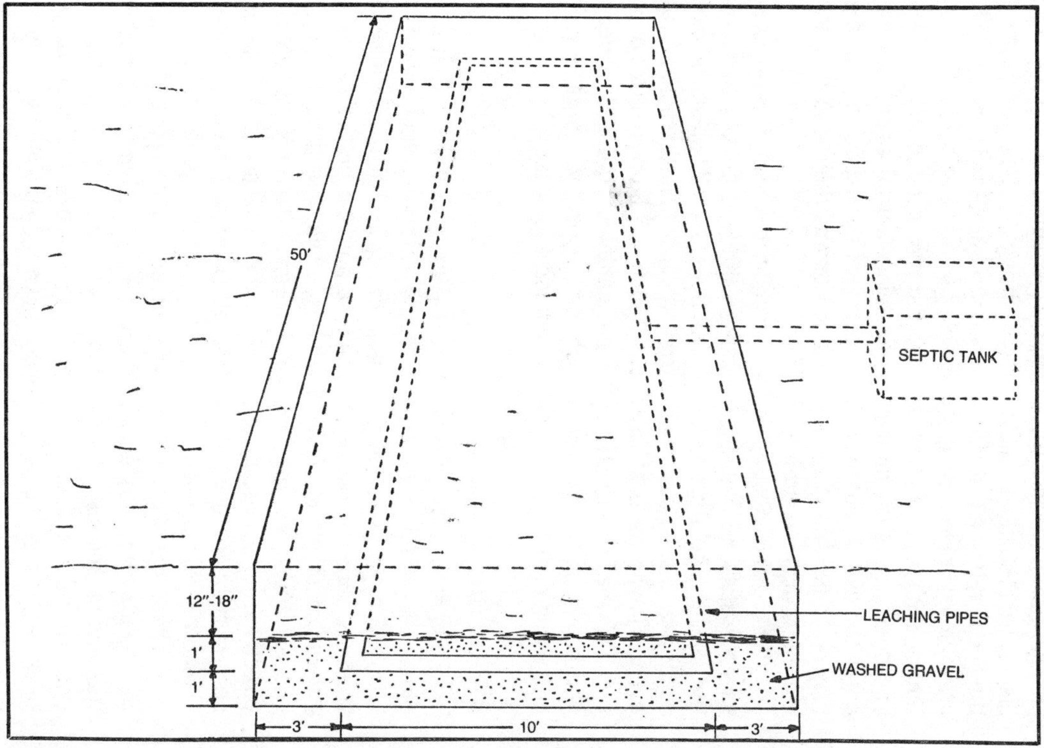

Fig. 7-6. Cutaway of typical seepage bed arrangement.

SEPTIC TANK

LEACHING PIPES

WASHED GRAVEL

3'

10'

3'

50'

12"-18"

1'

1'

necessary is to switch over at some appropriate point beyond the septic tank from solid pipe to be perforated pipe. The line, incidentally, need not necessarily be straight. In fact, single-line systems often cock off at one angle or another or are arranged to suit a sloping site, even to the extent of zigzagging back and forth along the contour lines.

However, if the leaching system consists of a number of drainpipes laid out in grid fashion and interconnected by a header and perhaps a tailpipe as well, another consideration may enter the picture. In some places running the sewer outlet line directly into a disposal field header by means of a tee fitting is not allowed. Instead, the line is run to a unit known as a *distribution box* (Fig. 7-8). The leaching field pipes are also connected to the distribution box, the theory being that effluent entering the box will exit more or less equally into all of the drainpipes. Each individual drainpipe leg may enter the distribution box. In some cases two or three main legs may be connected to the distribution box and then branch off into more legs further out into the field.

Where a distribution box is not required, usually it is unnecessary to install one. Running the sewer outlet line directly to a tee in the leaching field drainpipe header does a credible job in most circumstances. Another possibility, which does insure a somewhat evener flow, is to substitute a vee-branch or Y fitting in place of the tee. This may require the addition of some bend fittings in order to get the pipelines routed in the right directions, but that is no great chore and doesn't involve much additional expense.

BYPASS LINE

You have one more decision that must be made. That is to determine whether or not a gray-water bypass line should be installed in your system. This is a drainpipe that travels directly from the laundry area of the hours to a connection in the sewer outlet line or at the distribution box, bypassing the septic tank entirely (Fig. 7-9). The purpose of the line is to carry waste water directly from a washing machine to the leaching field. The line serves that appliance only, and no other drain. Here's the theory behind the arrangement.

There are two problems. One is that septic tanks and laundry detergents frequently do not get along well together. A big washing machine in daily use generates an incredible quantity of suds and detergent slime. These substances can raise havoc with the proper operation of a septic tank, slowing down or perhaps even stopping

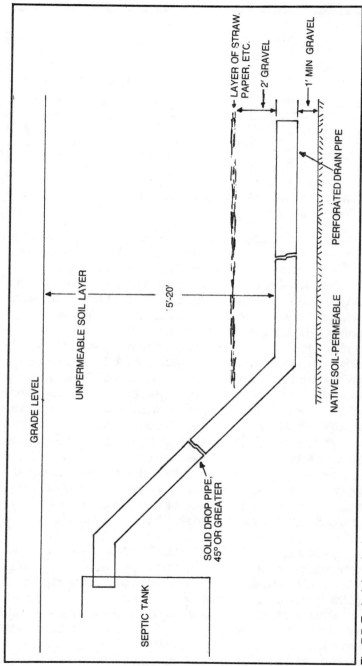

Fig. 7-7. Typical deep bed arrangement using 45 degree drop pipe inlet.

281

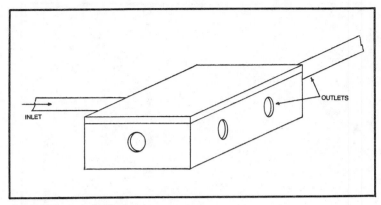

Fig. 7-8 Typical distribution box.

the bacterial action necessary to decompose the sewage in the tank. This situation, of course, can create a domestic disaster. When the tank stops working the users are immediately in trouble.

The second problem is simply that a frequently used washing machine runs a tremendous quantity of waste water into the septic tank, and usually at a rapid rate. This can have a couple of consequences. One is that the strong rush and outflow of liquid can carry a goodly portion of that all-important bacteria out into the leaching field where they can do no good whatsoever. At the same time, the decomposition capability of the tank is diminished. The action slows down and in severe cases may stop altogether. Another consequence is that raw sewage, stirred up by the inrush of waste water, can be carried directly into the leaching field before it has had a chance to decompose—an unlovely situation. Also, sludge or solid particles that have been stirred up can be carried out in to the field. These particles quickly plug up not only the microscopic air spaces in the way surrounding soil through which the effluent percolates away, but in severe cases can even plug the perforations in the drainpipe. The result is a completely inoperable drainage field that soons turns into a noisome marsh.

If you anticipate that the laundry facilities in your house will enjoy constant and heavy use, it might be best to install a bypass line of that is allowable under your local code. Incidentally, there is an alternative to making the drain connection in the leaching field itself. This is to route the line off in a different direction and run it directly into a dry well of substantial size. Like the leaching field, the dry well should be located where the percolation rate is as high as possible, and also away from boundaries, buildings and water sources.

RUNOFF WATER AND CHEMICALS

For many of the previously mentioned reasons, roof gutters, downspouts, storm sewers or any other outside drainage lines should never be piped into a septic tank. This would also include such items as garage floor drains or cellar drains. There is also one additional reason against this practice. Such drain water frequently carries quantities of silt and other materials that the septic tank can not digest. This will merely join the sludge at the bottom of the septic tank, and mean that the tank will have to be pumped out and cleaned far more frequently than would otherwise be necessary. If this type of drain water is routed into the leaching field, it will serve only to quickly plug up the drainlines. Dispose of runoff water by draining it into a ditch, dispersing it over a wide area to soak into the ground surface, or run it to a cleanable dry well built for the purpose.

One more point. If you should happen to be a photographer who enjoys doing the processing work in his own darkroom, by no means should you connect the darkroom sink drain to a house drain leading to a septic tank. Chemical substances, especially in the strong solutions that are regularly used in darkrooms, have the effect of killing off the bacteria in the septic tank even though the solutions may be quite diluted after entering the tank. The result is that the solids in the tank will not decompose and tank action slows down or stops. The darkroom drainpipe should be run to a dry well made for just this purpose. The well should be good-sized and fairly

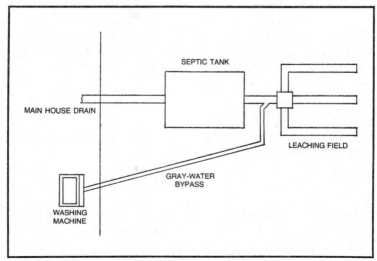

Fig. 7-9. Schematic of gray-water bypass system.

deep, since chemical solutions also have a deleterious effect upon plant life. Make sure that the drainage goes down deep where it can do no harm and will eventually be widely dispersed into underlying soil strata. And by the same token, never dump cleaners, thinners and similar chemicals into the house drain. This includes commercial drain cleaners; if a trap or a line becomes blocked, clean it out with a closet auger or snake.

DRAWINGS AND TAKEOFFS

Now you have all the information and details that you need to make up a drawing of the system. This need not be anything very fancy, and a line sketch with indications of all the necessary parts is sufficient. Dimensions may or may not be included, as you wish. The sketch in Fig. 7-10 is typical.

Follow your original block diagram for the general layout, and start at one end or the other of the system and line out the pipelines. As you do so, keep in mind the various direction changes that the line must make, and wherever a fitting is necessary indicate what it is on the drawing. When you are done, all of the bits and pieces that you will need to make up the complete system should be listed on the drawing, with the exception of miscellaneous supplies like pipe welding solvent.

From this drawing you can make up a *material takeoff*. This is nothing more than a list of all the parts, with their sizes and quantities, that you will need for the job. Where necessary, include product names or parts specifications as well. This will give you a complete grocery list to go shopping with. When you have listed all of the materials, you can also add any supplies and their approximate quantities that probably will be needed. If you wish, you can present a more formal typed copy to your plumbing supply dealer for a cost estimate or a firm quotation.

SEPTIC SYSTEM OPERATION

A septic system is by no means the best possible way for an individual homeowner to rid himself of sewage. That is better accomplished by disposal into a municipal sewage line. A septic system is not a particularly efficacious means of eliminating waste any way you care to look at it. But, sewage must be disposed of in some fashion, and this is the most universally accepted and commonly used way. There are better methods. All have such large drawbacks as far as the individual homeowner is concerned, though, that from a practical standpoint there is not better way. Until or

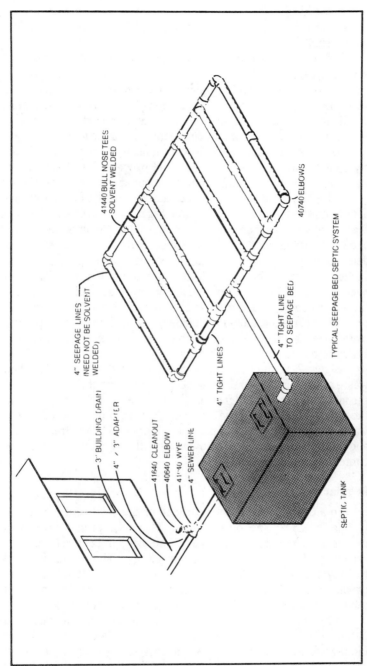

41440 BULL NOSE TEES
SOLVENT WELDED

40740 ELBOWS

4" SEEPAGE LINES
(NEED NOT BE SOLVENT WELDED)

4" TIGHT LINE
TO SEEPAGE BED

TYPICAL SEEPAGE BED SEPTIC SYSTEM

4" TIGHT LINES

3" BUILDING DRAIN

4" / 3" ADAPTER

41640 CLEANOUT

40640 ELBOW

41540 WYE

4" SEWER LINE

SEPTIC TANK

Fig. 7-10. Layout of typical septic system showing component parts (courtesy of Genova, Inc.).

285

unless those drawbacks are overcome, septic systems will continue to be installed.

But this system can be a tricky one. The installation is disarmingly simple, and once installed it is largely out of sight and out of mind, at least until something goes wrong. This is an all too frequent occurence. Since the operation of a septic system is so little understood and since the system is so frequently abused, a few words about proper operation and maintenance are in order.

The operation of a septic tank is simple in concept, yet incredibly complex in the doing. To start the process, liquid nonhuman waste is joined by both liquid and solid human waste in the house drain. In addition, there may be a certain amount of grease, particles of grit and sand, animal fat and bits of vegetable matter and the like. If a garbage disposal unit is present, considerable amounts of garbage are added to the mix. Because garbage places an added demand on a septic tank, in some places garbage disposal units are forbidden. On the other hand, in other areas their installation is encouraged. All of the material that ends up in the house drain, with the exception of silt and grit particles (from floor washing, etc.) is, or at least certainly should be, biodegradable. This means that the material is capable of being broken down or decomposed in a fairly short time without much difficulty.

Figure 7-11 shows the makeup of an operating septic tank. The waste passes through the house sewer and enters the septic tank at the inlet. Here a *baffle plate* keeps the waste from shooting out into the tank and disturbing the surface. The material is routed downward to join the contents of the tank a foot or so below the surface, and mixes in at a relatively slow rate with little disturbance of the contents already present.

Putrefaction

In a continuous action, nearly all of the solid contents of the tank are undergoing putrefaction. This is a process whereby *anaerobic* (active in the absence of free oxygen) bacteria and fungi are busily chewing up the waste and splitting the proteins. Most of the material is liquified and composed of incompletely oxidized products of various sorts with a characteristic foul or sickish-sweet odor. Certain enzymes are also at work to aid the process.

The heavy materials settle on the bottom of the tank in a nonuniform layer, and those that are not readily decomposable or biodegradable may remain there indefinitely. A certain amount of this *sludge*, as it is called, will never decompose and slowly builds

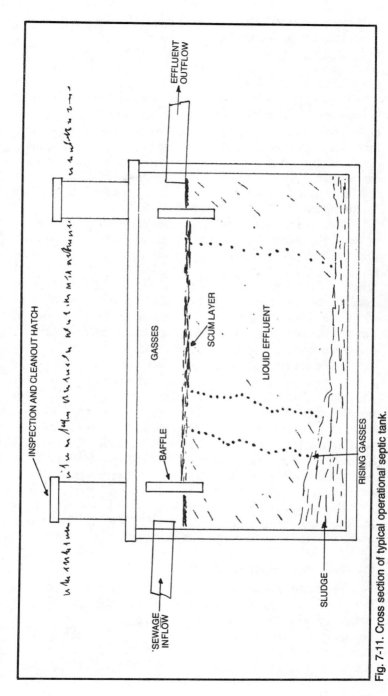

Fig. 7-11. Cross section of typical operational septic tank.

287

up. Greases and any materials that are lighter than the liquid content of the tank rise to the surface to for a *scum* or *crust*. The composition and thickness of this crust depends largely upon the kind of waste material being run into the tank and the effectiveness of the decomposition process taking place.

The liquid between the sludge and the scum is called *effluent*. It is usually cleanest (if that word can be used) and clearest at the upper level on the far side of the tank. This is where the outlet lies, draining into the sewage outflow line and then to the leaching field. As waste runs in at the tank inlet, an identical amount of effluent drains out into the leaching field. Despite the purtefaction process, this liquid is still sewage in essence and presents·an enormorous health hazard. It is not purified, nor even close.

The effluent enters the drainpipes in the leaching field in more or less equal fashion for all pipes, and slowly bleeds away through the pipe perforations and into the soil. Meantime, *aerobic* bacteria are at work here in the process of oxidizing waste products and breaking them down into harmless substances. This process continues as the effluent seeps in to the soil. After a long enough period of time elapses and after the effluent has traveled far enough through the soil, it eventually does become purified. In fact, its pretty much loses its original characteristic altogether.

As the purtrifying process takes place within the septic tank, a certain amount of gases are released. The gases originate in the depths of the tank contents. By bubbling up through the liquid they occasion very mild agitation of the effluent, which in turn increases bacterial activity in the putrefaction process. The result is that most solids in a properly operating tank are digested in a matter of 24 hours or so. The gases, which can be highly toxic, explosive or both, find their way out of the tank through the venting portion of the drainage system.

The scum that forms on the surface of the effluent also helps in the decomposition process. It has the added effect of sealing in odors and preventing any aeration of the contents.

Problems Retarding Bacterial Action

The bacterial action of a septic tank is what makes the whole thing work. Though the system will operate unfailingly under the right conditions, the delicate balance can easily be upset through improper use. Nothing need be done to a new tank save to make sure that is cleaned of debris when installed. The operation is self-starting. By the time the tank is full for the first time and ready to discharge effluent, bacterial action will be well under way. But a number of circumstances can retard or even destroy bacterial action.

Too little use can sometimes be a problem. Overuse is another, where the influx of raw sewage exceeds the decomposition capabilities in a 24-hour period. An overabundance of gray water and very little solid material, or vice versa, can also cause problems. Sudden surges of large quantities of waste being discharged into the tank can roll up the contents, slowing decomposition and quite possibly forcing solid particles into the leaching field to plug up the lines. Water surges or a lack of proper baffling can disturb the surface scum, rendering it partially ineffective.

Material that will not decompose in a tank within 24 hours or so can be considered nonbiodegradable for these purposes, even though that same material might eventually disintegrate under other circumstances. This means that newspaper, paper towels, paper napkins, cloth or any similar material should never be disposed of in a septic system. Neither, of course, should hard or solid material. Even if such items do manage to get into the tank without plugging up the lines partially or completely, they will remain there indefinitely. They will not decompose, and will only add to the sludge layer on the bottom of the tank. After that sludge layer reaches a certain depth and the layer of liquid effluent above it diminishes to a certain depth, the tank will become inoperative. It must be pumped out, the contents take away and tank operation begun once again. This process is both inconvenient and expensive.

The presence of chemicals, cleaners, paints or paint thinners commerical drain-cleaning fluids, concentrations of detergents and the like also can have a drastic effect on the tank operation. An operating septic tank generates its own heat, and is unlikely to freeze. However, the cooler the contents become the less effective is the bacterial action. In areas of severe winter weather a tank should be buried somewhat deeper than normal, perhaps 2 feet below the surface.

Inspecting the System

In order to maintain effective septic tank operation with a minimum amount of trouble, the system owner must make a point to inspect the system regularly. He must take pains that it is properly used by himself and the other occupants of the building as well. Proper use of the system is largely a matter of common sense. If the system is correctly installed to begin with, it will perform nicely for a couple of decades or more. There are a number of things that can be checked on a regular basis.

The presence of an obnoxious odor near the tank can mean that it is not operating properly. If the soil around the leaching field becomes saturated, obviously there is a problem. A dry leaching field combined with soggy ground around the septic tank may mean a blockage. You can periodically inspect the contents of the tank through the inspection hatch, too. If the scum layer seems inordinately thick, steps should be taked to reduce it before it clogs the outlet pipe or moves out into the field. You can take regular measurements of the sluge depth on the bottom of the tank by using a graduated stick. At some point the increasing depth of the sludge and the decreasing depth of the effluent layer will commence to reduce the effective operation of the tank. This is the time to have the tank pumped out before the system clogs up completely. The supplier of the tank or your local septic tank service company will be able to give you some idea of what that point might be for your particular size and configuration of tank.

In addition to the factors just discussed, there is another way to help your septic tank maintain constant and effective operation. Do it by the addition of a special compound called a bacteria additive, poured into the tank at periodic intervals. The object of these compounds is to encourage and increase the activity of bacteria and enzymes in the tank to decompose the solids more quickly and effectively. They will give you a boost to a slow-operating tank and frequently bring a nearly defunct one back to life. Just how beneficial the addition of such compounds might be in any given case is difficult to determine. There is no means of comparing with another identical tank used under identical circumstances but without the addition of a compound. However, the whole idea is a good one and in most cases the compound undoubtedly helps. It's good insurance against a malfunctioning tank, at very low cost and little effort.

There are numerous brand names of these products on the market, such as Rid-X, Septic-Kleen and others. They are readily available at hardware stores and most supermarkets. Most of them do a particularly good job in breaking down fats, greases and oils so that they can be flushed away and not clog the system. The compounds are harmless, contain no harsh chemicals, are nontoxic and noncorrosive, and will not bother any part of the plumbing system at all. The usual application process is simply to flush a certain amount of the compound, depending upon the size of the tank, down a toilet once a month. That's all there is to it.

Another old trick that has been used for many years and is claimed by a good many folks to be extremely effective is the addition of yeast to the tank. The usual combination is a half-pound of Brewer's Yeast Powder mixed in about a gallon of hot water. This solution is poured into the toilet closest to the septic tank. After the water in the toilet bowl has resumed its normal level, the toilet flushed and the remainder of the solution sent into the system. The theory is that the yeast improves bacterial action in the tank.

Chapter 8

Sewage System Installation

Installing a septic system is not a difficult chore, aside from the physical labor involved, but it must be done right the first time around if you except to have a fully functional system. Making changes after the system is finished can be done, of course, but that's an awfully frustrating and expensive business. By following a few simple guidelines, though, you should experience no difficulties.

EXCAVATION

The installation of a septic system begins with making a series of excavations to accomodate the various components. The first step is to lay out the outlines of the excavations exactly where they must be made, using your sketches and drawings as a guide. Line up the house drain with the house sewer line, measure it out and drive stakes to mark the position of the trench. Then outline the pit for the septic tank, making the dimensions a couple of feet larger all around than those of the tank itself, in order to provide some working room. Line out the sewage outflow line trench next, followed by the leaching field. If the leaching field is a subsurface-tile system, you may have to set out stakes defining the outlines of one trench or several. In the case of a seepage pit or a deep bed, all you'll need is the outline of a rectangle of suitable dimensions. The excavator will then know exactly where he is supposed to dig. If your layout is in a hayfield or some other area of heavy ground cover, flag the stakes with strips of bright-colored plastic surveyors' tape for good visibility.

Potential Problems

After the layout is made, check over the surrounding area to see what kind of problems might arise from trees or shrubbery. A subsurface-tile leaching system should be exposed to as much sunlight as possible, and not be shaded to any appreciable degree by trees or high shrubbery may be necessary.

Roots can be even more of a problem. Trees growing near the pipeline can eventually displace or even break up sewer pipes or drainpipes in a leaching field. Some shrubs with extensive and vigorous root systems that lie near the surface can be much worse than trees. Lilac bushes are a notably good example of this; their roots will grow right into a pipeline through perforations or tiny cracks in the joints and literally fill the pipe, choking it off completely. Removal of such nearby shrubbery may be necessary. In some cases, such as the lilacs, removal will not be enough, since any missed live roots will continue to regenerate themselves. Treatment with an effective herbicide is often necessary. Similar problems can be encountered with seepage pits, but deep beds are seldom bothered because they lie so far beneath the surface of the ground.

The excavations can be, and often have been, done by hand. But it's a whole lot easier and faster, though more expensive, to hire an excavating contracto. He can do the whole job in a few hours with a back hoe or some similar machine. It's helpful to hire someone who has done this kind of work before and has some knowledge of septic tank installation. In any event, look for someone who has a good reputation, knows how to handle his machine and can do the job correctly and expeditiously. You might want to solicit two or three bids for the job, but many contractors prefer to do small jobs of this sort on the basis of an estimate, and then charge you for the time actually spent. Frequently this is determined rather exactly by a running-time or engine-hours meter mounted on the machine itself.

Proper Relationships

The first critical point in the excavation lies in the proper relationship of the house sewer trench bottom to the septic tank pit bottom. The house sewer line will come through the foundation wall, or extend from beneath the house, at a certain downward slope or pitch. The downward pitch must be continued in the sewer line all the way to the inlet of the septic tank. Consequently, the tank

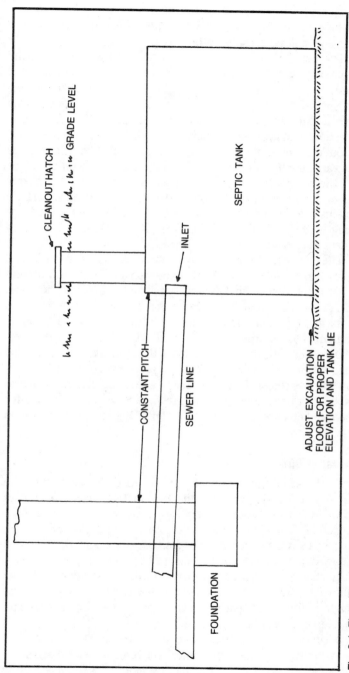

Fig. 8-1. Elevation of septic tank must be adjusted to exactly align with house foundation, house main drain, sewer pipe inlet and outlet, and approximate final grade in tank area.

must be positioned at just the right depth so that the inlet and the sewer line will meet properly (Fig. 8-1).

The relationship of the septic tank outlet to the lie of the sewage outflow line and the leaching field is not as critical, but is of consequence. Like the house sewer line, the sewage outflow line must pitch downward and away from the tank to a slight degree before it enters the drainpipe in the leaching field. Thus, that connection must be made at a point somewhat lower than the septic tank outlet. The drainpipes in the leaching fields must also pitch downward slightly throughout their entire length. Yet at the same time, the drainpipe must be set reasonably close to the ground surface. In many cases the ground contours are such that no problems are encountered. In others, a certain amount of recontouring in the leaching field area may be necessary in order to maintain the reasonable pipe depth.

The trench for the house sewer line should be dug 6 to 8 inches deeper than the level at which the pipe will lie. The trench bottom should roughly follow the downward pitch of the pipeline. The septic tank pit is best dug to as exact depth as is conveniently possible, so that its great weight will rest upon a minimum of loose fill and a maximum of undisturbed native subsoil. The trenches for the sewage outflow line and for the drainpipe in the leaching field should also be dug 6 to 8 inches deeper than the pipeline depth, and roughly follow the desired pitch. In the case of a leaching field made in grid fashion with a header and perhaps a tailpipe as well, the trenches for those two lines are dug to suitable depth, but level and having no pitch.

Disposal of Dirt

At the same time arrangements are made for excavation, be sure to give consideration to the fact that a certain amount of spoil—the dirt that comes out of the excavation—will have to be disposed of. The septic tank will displace a large amount of spoil, with only a small amount being packed back in around the tank wall and covering the top to a depth of from 1 to 2 feet, in most cases. A good portion of the spoil removed from the house sewer and sewage outflow trenches will be used for backfill. However, much of that taken from the leaching field trenches, or from either a seepage pit or a deep bed, will be replaced with gravel.

All in all, there is a good deal of dirt to be done away with. The least expensive and sometimes most convenient method of disposing of spoil is to work out ahead of time some types of recontouring or filling projects close by the construction site. This has the advantage of taking care of two projects at one stroke, at reduced expense. The alternative is to simply have the stuff trucked away. Wherever possible, though, try to save the topsoil, as this will make revegetation around the site much easier later on.

SETTING THE TANK

The usual procedure in making a septic system installation, after the excavation has been completed, is to set the septic tank in place first, properly aligned, and then run the pipeline to and from the tank. In some cases this may consist of actually constructing the tank in place in the pit. Some lightweight tanks can be muscled into place by two or three workmen, while others must be lowered with a backhoe boom or from a special truck designed to do just that.

Working on the Pit and Lid

In any case, the bottom of the pit must be leveled and adjusted to a suitable depth so that the tank inlet is properly aligned with the house drain, and the tank outlet aligned with the outflow line. The tank must also set level; otherwise there could be problems with proper inflow and discharge. And though mention of the fact may seem strange, it's also necessary to make sure that the tank is placed with the inlet facing the house and outlet facing the leaching field. In most tanks the inlet is positioned slightly higher than the outlet, so that it will not be submerged and the free flow of incoming sewage consequently hindered. If the tank is installed backwards, this is exactly what will happen. The proper pipe alignment will also be thrown out of whack. Surprisingly enough, this does occur. With the round steel tanks, for instance, it is difficult to tell at a glance which port is the inlet and which is the outlet. By placing some sort of unmistakable identification on the inlet port, the problem can be avoided. Some tanks, of course, obviously can be placed in only one position.

Once the tank is positioned and leveled, if there is no reason for any work to be done from the inside of the tank (making connections, etc.), the lid should be set in place. This prevents dirt and debris from getting into the tank and also reduces the possibility of injury from someone slipping and falling in. The covers for concrete tanks must generally be set in place by machine, but steel tank covers can be dropped on by hand. This is the time, too, to consider access or inspection hatches. Many tanks come so equipped with a small removable lid of some sort placed above the inlet port, and often above the outlet port as well. These hatches are sometimes made flush with or just above the cover surface. When the tank is buried, so are the hatches. They must be dug up every time an inspection is necessary. Other tanks are equipped with extension hatches, so that the covers lie just above ground level; this is by far the better approach.

Inspection Hatch

Some tanks, notably the inexpensive steel ones, are not provided with inspection hatches at all. If this is true in your case, by all means take the time to make up at least one extension hatch and mount it on the cover over the tank inlet port. There are a number of simple ways to go about it. About the easiest is to make a box column of construction-grade redwood 2 × 10s or 2 × 12s in an appropriate height, say 2 feet or so, or whatever will bring the top of the column about 6 inches above ground level. Cut another piece of redwood, or heavy sheet metal, to fit atop the column as a lid. The lid can be secured with four short lag screws.

Cut a hole that matches the inside dimensions of the column into the tank cover. If the cover is steel, the column can be affixed to the cover with a couple of screweyes, steel angle brackets or corner braces screwed to the column and bolted to the cover. Applying a coating of mastic, roofing cement or some other suitable compound to the bottom of the column to seal it to the cover (Fig. 8-2). The same method can be used with a concrete cover, by lagging the brackets down with shields and bolts. The column can be set in an extra-thick bed of heavy mastic and held securely in place by hand as the backfill is placed. Once the fill has settled and compacted, the hatch is unlikely to move.

Alternative methods that work well are to use a length of vitreous clay drainpipe or concrete tile of suitable diameter. These can be set in mastic atop a steel cover, or mortared in place atop a concrete one. You might use lengths of large-diameter heavy steel pipe or sections of culvert. Again, these must be securely sealed to the cover opening. Lids for these access hatches can be made of precast concrete of somewhat larger diameter than the outside diameter of the pipe and perhaps 3 or 4 inches thick. The weight of the concrete lid will hold it in place.

When working with any septic tank whose sides are a bit flexible, such as a steel tank, it is important that no backfilling be done around the tank until the lid has been put on. It's a good idea to backfill here as soon as is reasonably convenient because of the

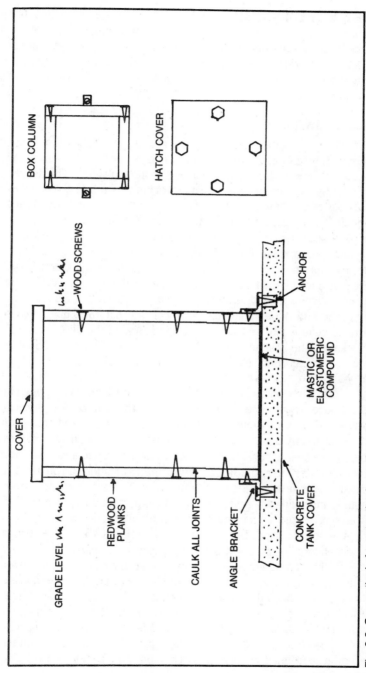

BOX COLUMN

HATCH COVER

WOOD SCREWS

COVER

GRADE LEVEL

REDWOOD PLANKS

CAULK ALL JOINTS

ANGLE BRACKET

CONCRETE TANK COVER

MASTIC OR ELASTOMERIC COMPOUND

ANCHOR

Fig. 8-2. One method of constructing inspection and cleanout hatch for septic tank.

safety hazard presented by the open pit, but haste in this instance cause problems. The difficulty is that the uneven pressure of the backfill throws the tank walls out of skew so that the cover will not fit. The usual result is that the tank has to be at least partially dug up again and the pressure on the sides relieved before the cover can be fitted.

HOUSE SEWER LINE

With the septic tank set in place and properly aligned, you can turn your attention to installing the house sewer line. As mentioned earlier, this line must pitch downward to the tank inlet. The degree of pitch is slightly variable, but not very much so. The pipe sizes used for plastic house sewer lines are usually 4-inch and sometimes 3-inch. The minimum pitch for these pipes is ⅛ inch per running foot of pipe, and the maximum allowable pitch is usually considered to be ½ inch per running foot. If the pitch is less than ⅛-inch, liquid drainage is slow and is also insufficient to move solid material down the pipe and into the tank. The inevitable result is clogging of the line after a short period of time. If the pitch is greater than ½-inch, liquid flow is quite rapid, but so much so that the liquid rushes around solid waste and leaves it behind. Again, the likely result is blockage. The best and most often used compromise is a pitch of ¼ inch to ⅜ inch per running foot of pipeline. The ¼-inch pitch is probably the most frequently used.

There are occasions where running a consistent slight pitch in the house sewer line from one end to the other is a physical impossibility. This would be the case, for instance, when because of the topography the septic tank inlet must be positioned several feet below the house foundation but only a short distance away. In such cases an alternative method of sewer line installation can be used. If a soil pipe or sewer line is set at an angle of 45 degrees or more, both liquids and solids will zip right down the pipe with no problems. The 45-degree angle is a minimum, and any other practical angle up to 90 degrees can be used. But because of the fact that 45-degree fittings are standard, using them is the most practical solution.

If this arrangement is necessary, start laying the house sewer line by joining a section of pipe to the house drain. Lay the pipeline toward the tank in the usual manner with a ¼-inch downward pitch. At the proper point, which you can determine by measurements, attach a 45-degree fitting and continue the sewer line downward on a 45-degree angle. At the next appropriate point, attach another 45-degree fitting in reverse position, along with a short stub of pipe to connect to the tank (Fig. 8-3).

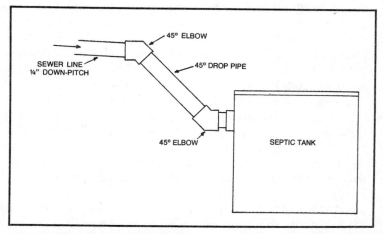

Fig. 8-3. Where constant pitch of sewer line cannot be maintained, final section of line can be installed as a 45 degree drop pipe.

Setting the Grade Board

The first step in preparing to lay the pipe is to clear out the trench bottom so that you can work in it easily. The next is to set a *grade board* the full length of the trench, angled to a continuous ¼-inch downward pitch. The grade board can be an inexpensive grade of ordinary 1 × 6 or 2 × 6 wood stock. Practically any kind will serve, though redwood has considerably greater longevity (and costs more) than pine, spruce or fir.

The grade board is set on edge and nailed to stakes driven into the trench bottom (Fig. 8-4). The top edge of the grade board

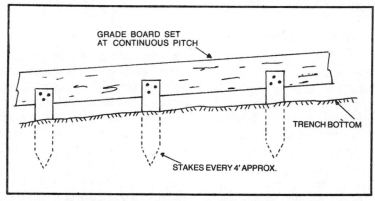

Fig. 8-4. Grade board staked into trench bottom assures ample sewer pipe support, constant pitch of line and eliminates sagging.

should lie at exactly the same height as the pipeline bottom. The pitch must be continuous with no bellies or humps, so it's best to choose planks that are not bowed. The grade board is installed beneath the pipeline in order to keep it from settling and sagging, or possible cracking open, as the trench backfill settles and compacts. No grade board is needed, incidentally, under sewer lines running at a 45-degree pitch—only under those with a gentle pitch.

Once the entire run of grade board is installed, double check to make sure that you have the right degree of slope. This can be done by measuring the drop from one end to the other of the pipeline. Attach a string to the top of the pipe where it joins the house drain, and stretch the line taut and level to the other of the pipeline. Attach a string to the top of the pipe where it joins the house drain, and stretch the line taut and level to a point directly above the other end of the pipeline. You can check for level by attaching a line spirit level to the string. Then measure the distance from the string to the top of the outlet end of the pipe. In a 20-foot run of ¼-inch pitch, for instance, the total drop from one end to the other should be 20/4 inches, or 5 inches. If the drop is correct, lower the taut line until it touches the top of the grade board. Sight along the string to make sure that the top of the grade board is indeed flat for its entire length and does not bow down or hump up. (Fig. 8-5).

The next step is to pack backfill all around the grade board. Select the dirt carefully, and make sure that it is free from debris and rocks any larger than about golf ball size. If the spoil dirt is extremely rocky, you may have to import some clean fill to work with. Fine gravel will also do the job, and in fact is better. Bed the fill in and tamp it down firmly, but not so hard that you disturb the lie of the grade board.

Setting the Pipeline

Now the pipeline can be set in place, and there are two ways to go about this. One is to join the first section to the house drain, and then keep going section by section and fitting by fitting until you arrive at the tank. The other possibility is to make up the bulk of the pipeline out of the trench and on a relatively flat surface where the working is easier. Then the whole pipeline can be lowered into the trench and connections made at the house drain and the septic tank inlet.

This pipeline must be of solid pipe connected with tight and sealed fittings, so that no raw sewage can possibly escape into the surrounding soil. Some types of pipe are put together with coupling

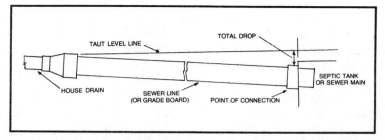

Fig. 8-5. Method of checking for proper pipeline drop from end to end, as well as constancy of pitch.

and fittings solvent-welded to make the joint. If this kind of line is put together as a unit outside the trench, it should be allowed to cure overnight before the pipeline is manhandled into the trench. This allows the joints time to gain full strength and avoid the possibility of a weak or leaky jointure. If the pipeline is made up in the trench where it will not be moved around, there is no problem and back filling can begin immediately. Other types of pipe, like Genova's "Liberty Bell," use a push-coupling system and go together "dry." Just shove the sections together in the trench (or out on the ground surface) and the job is done (Fig. 8-6).

Line the pipe up carefully so that it lies exactly atop the grade board, and shovel clean fill around the sides of the pipeline. Pack the fill down firmly and carefully, keeping an eye out that the correct lie of the pipeline is not disturbed. Follow up by shoveling in more fill to a depth of 3 or 4 inches over the pipeline, but don't tamp. You might damage the pipe. Then the remainder of the trench can be backfilled either by hand or machine. It's an excellent idea, though, to refrain from driving a heavy backhoe or crawler tractor over the trench as there is a danger of crushing the line. Pack the backfill down reasonably firmly with a hand tamper, and leave the surface humped up a few inches above the surrounding grade level. As the soil settles and compacts, the filled area will eventually more or less equalize with the surrounding terrain.

Making the Connections

There is sometimes some confusion as to how the house sewer line connections should be made. Connecting to the house drain should present no problem. In some cases it is permissible (as well as easy and advantageous) to simply continue the line directly to the septic tank using the same kind of pipe and fittings. Where the house drain is of 3-inch diameter and the house sewer line 4-inch diameter an adapter fitting makes the transition easily. Disparate types of pipe, such as a PVC house drain and an ABS sewer line, can be joined by solvent-welding in the usual fashion. If the pipe types are unlike or are mechanically joined, all that is needed is the correct adapter.

Connection at the septic tank inlet can sometimes pose some questions. Some tanks are provided with only inlet and outlet holes or ports, with no connection fittings or pipes whatsoever. The holes are almost always larger than necessary for plastic pipe. There are a couple of possibilities here, and exactly what you do depends on how the tank is made.

If concrete, just run the end of the house sewer pipe through the port until it sticks out inside the tank about half an inch. Seal the pipe in place with mortar packed into the excess space and built up in a smoothed out ridge around the pipe on both the inside and outside of the tank.

If the tank is steel, you may have to first locate an adapter fitting the large end of which completely or nearly fills the hole in the tank. Join this fitting to the end of the sewer pipeline so that it extends about half an inch on the inside of the tank. Join this fitting to the end of the sewer pipeline so that it extends about half an inch on the inside of the tank. If there is a lipped flange around the hole, the entire void can be filled with plastic cement, oakum and lead wool, an elastomeric compound or some similar sealant. If there is no flange, the pipe end will have to be sealed into the hole by whatever means seems to be most advisable and will hopefully make an effective seal. This will probably mean heavy applications of a suitable sticky and non-hardening sealing compound that can be laid on and built up in a tapered ring around the pipe and against the tank wall.

Where the tank is provided with pipe stubs or connection fittings, they will probably be made of clay tile or cast iron. In either case your problems are somewhat lessened, because adapters are available to connect various types of plastic pipe to either clay or cast iron. Generally these connections are made by setting the pipe and fitting together and carefully caulking them with a sealing compound.

However the connection is made and whatever the kind of pipe or fitting is used, liquid leakage is not likely to be a serious problem. This is true because the pipeline runs into and beyond the tank wall at a downward pitch, so waste traveling through the pipeline will exit without coming into contact with the connection seal. There is no pressure involved, and if the tank operates correctly there will be

Fig. 8-6. Sewer line being set in trench.

305

no backup of effluent to flood the inlet area. The seal, then, has its primary function in preventing the admittance of outside dirt and ground moisture, containing minor amounts of splashback and confining gases to the interior of the system. Nor will the seal have to withstand any mechanical strain, assuming that the house sewer line is correctly installed and does not settle or pull back. Even a relatively light and fragile seal should hold perfectly well for an indefinite period of time.

SEWER OUTFLOW LINE

The sewer outflow line is connected to the outlet of the septic tank and extends to the starting point of the leaching field drainpipe system. Like the house sewer line, this discharge line must be tightly sealed at the tank outlet and solidly jointed throughout its length, so that no effluent can escape into the surrounding soil until it reaches the leaching field. The line should be laid at a ¼-inch to a ⅜-inch downward pitch to ensure good effluent flow.

Often it is possible to join the septic tank to the leaching drainpipe by means of a single 8-foot or 10-foot length of pipe, or the distance can be even shorter. The shortest practical distance is generally considered the best, especially in areas of severe winter weather where ground frost is likely to go deep. There is always the possibility in a long sewer outflow line that small discharges of effluent into the pipe will freeze before they reach the leaching field, slowly building up an accumulation of ice in the line that will eventually choke the pipe off completely. Freezing is frequently less of a problem with the house sewer line, because there is an opportunity to bury it deeper. The average temperature of the discharges into the pipe is likely to be higher. If freezeup is a potential problem in your area, ask around to see what the local practice is in making septic system installations that won't freeze.

Installation of the sewage outflow line is made in just the same way as the house sewer line. The exception is that if the pipeline is very short and the pipe itself reasonably rigid (or in the case of a 45-degree drop outlet), the grade board can be eliminated. The pipeline in that case is best laid upon solid, undisturbed subsoil, carefully smoothed to the proper pitch and cleaned free of rock or debris. If loose fill must be added to bring the pipeline into proper position, make sure it is tamped down hard.

The pipe used for the sewer outflow line is generally the same as that used for the house sewer line. Fittings or other pipe jointures are made in the same manner. The connection to the septic tank outlet is made in the same way as the inlet.

LEACHING PIPE

The last part of the operation involves the installation of the leaching pipes or absorption field drainpipes. There are three basic possibilities here. One is a straight-line continuation of the sewer outflow line as a drainpipe, and to this might be added the possibility of one or two Y-branches in order to follow ground contours. The second possibility is a grid system, utilizing tees and elbows. The third involves installing a distribution box as a nucleus.

In a single-line drainpipe, or some close derivation thereof, the first step is to drive stakes and set a grade board the full length of the trench, just as you did for the house sewer line. In this case, however, the pitching arrangement is slightly different, and may be regulated by local code. The usual arrangement is to pitch the drainpipes to a downward slope, away from the septic tank, at a rate of from 2 inches to 4 inches per 100 feet of run. This amount is just a shade less than ¼ inch to ½ inch for every 10 feet of pipe. Generally, the steeper angle is the preferred one.

However, if the pipeline actually does extend 75 to 100 feet, a slightly different system is used. The first 50 feet of pipeline is

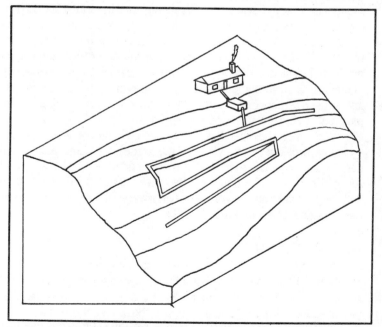

Fig. 8-7. Leaching line installed at continuous pitch along contour lines of slope and jumping from level to level with drop pipes.

pitched at a rate of approximately 4 inches per 100 feet. A single-line branching system that follows ground contour is often pitched at 4 to 5 inches per 100 feet everywhere. Note, however, that these figures do not always hold true. In the absence of any regulations to the contrary, they have proven to give satisfactory operation. But some knowledgeable folks recommend a constant pitch of about 12 inches per 100 running feet, while other authorities insist that the drainpipe be laid perfectly level throughout. Also, in some locales the final 15 or 20 feet of each drainpipe must be pitched slightly upward, to avoid any possibility of pooling of the effluent at the end of a line.

Pipelines that follow contours and jump from grade level to grade level are generally laid out with the branches going with the contour and a slight downward pitch. The interconnecting lines between branches are set at a relatively steep angle, dropping sharply down to successively lower levels (Fig. 8-7).

Graveling the Trench

Once the grade boards are installed at the proper pitch and check for straightness, the trench is filled with gravel. Requirements here vary considerably. Washed gravel in ¾-inch size is perhaps most often used. However, local regulations may say otherwise, perhaps calling for a graded run of gravel ranging from ½-inch to 2 ½-inch size. There may be no requirement for the gravel being washed. This is the best bet since washing eliminates the possibility of quantities of silt being sluiced off the gravel by the drain effluent, and subsequently reducing the porosity of the surrounding soil by plugging it up with tiny mineral particles.

Whatever the specifics, the gravel should be laid into the trenches, worked carefully around the grade boards and stakes, and packed in firmly. The top of the gravel should be level with the top of the grade board. The depth of the gravel at this point should be 6 inches at a minimum, covering the entire trench bottom. A greater depth is fine and may be required.

Laying the Pipeline

With these chores out of the way, the pipeline itself can be laid. Perforated plastic pipe is the kind to use, in a 4-inch diameter and preferably of a type that matches the sewer outflow line. The perforations are placed downward, and the first section is coupled to the sewer outflow line with the standard coupling fitting for that kind of pipe. Subsequent pipe sections are successively laid down the length of the trench, resting directly atop the grade board. Couplings can be made dry; that is solvent-welding is not necessary here. Just push the sections together.

Now comes more gravel. Again, the amount necessary may be regulated by local code, but the top of the pipeline should be covered by a minimum of 2 inches of gravel. More can be added, up to a depth of 3 or 4 inches above the top of the pipeline, but beyond that little purpose will be served. Make no attempt to compact this gravel, as there is too much danger of breaking or crushing the pipe or disturbing its lie. Just rake out the surface so that the gravel is fairly level.

An interface of some sort must be laid atop the gravel to prevent the backfill dirt from filtering down into the gravel and clogging the leaching area. Ordinary building paper, obtainable at lumberyards and building supply houses, is a common choice. The material must be permeable to moisture, however, so that proper evaporation can take place from the field surface. This means that plastic film or sheeting, tarpaper or roofing felt, which are essentially waterproof, cannot be used. Another possibility is to put down a 2-inch layer of straw (not hay). Whatever material is used should extend from side to side in the trench and cover the entire pipeline.

Once this is done, backfilling can begin. The backfill should be clean and free of debris and large rocks. Do the first part of the job by hand, shoveling fine dirt carefully onto the interface so as not to disturb or puncture it. The remainder of the backfilling can be done either by hand or by machine, but in the latter case considerable caution should be exercised so that the trench sides are not caved in. A heavy machine should never be run across the trench. Don't attempt to compact the soil, but merely have it heaped up so that it can settle out of its own accord. As mentioned earlier, there will be a considerable amount of excess spoil which must be disposed of. Figure 8-8 shows a typical completed installation.

Installation With a Distribution Box

An installation involving a distribution box is done in essentially the same manner, and the differences lie only in a few specific details. Here, the first step is to set the distribution box in place, preferably upon a level pad of undisturbed subsoil so that the box won't have a tendency to settle into the ground. The box should also be as level as possible. At the same time, connect the sewer outflow line to the distribution box, using whatever fittings or methods of caulking and sealing are indicated.

The grade boards can next be established in the pipeline trenches that lead back to the distribution box. They should be adjusted to match the level of the distribution box outlets, and be pitched downward from that point as necessary.

Laying the drainpipe lines is a matter of connecting the first sections to the distribution box outlets, and then continuing to add on the remaining sections until the lines are completed. The rest of the installation is just the same as for the single-line arrangement just discussed.

Installing a Grid Network of Pipes

Installing a grid-type network of drainpipes poses a slightly different situation. The first step is to attach a tee fitting to the end of the sewer outflow line, at the point where the leaching field begins. The tee should be solvent-welded or otherwise solidly secured in a leakproof fashion, with the crossarm of the tee set exactly level.

A short stub of pipe is then connected to each side of the tee crossarm, to extend the required distance to the next tee. This process is continued until there is one tee in place for all but two of the required number of drainpipes. The remaining two are the ones located at each end of the grid, and a 90-degree elbow is installed at each location. All of these fittings should be made up tight and leakproof, and lie on a level line. If this header assembly is of any appreciable length, you might wish to set it upon a grade board to make sure that it remains level. Gravel is not necessary beneath the header as this is not part of the leaching field proper, but rather an effluent distribution line. The line can rest upon undisturbed subsoil if a grade board is not desired, and spoil dirt used for close fill around the pipe.

The next step is to make ready for the drainpipes. If they are relatively short, it may not be necessary to rest them upon a grade board. However, if they are longer than about one length of pipe, grade boards are a good idea in order to maintain proper alignment and pitch and to avoid the possibility of sagging when the backfill settles. Set up the grade boards if necessary, and/or spread a layer of gravel in the trench bottom, just as with a single-line system. Double check to make sure that the pitch is correct.

The perforated drainpipes, with the holes facing downward, can now be laid by simply shoving them into the tees and elbows. Leakproof joints are not necessary here. If the drainpipes are two or more sections of pipe in length, they can be coupled together with a push-fit; these joints do not need to be leakproof, either.

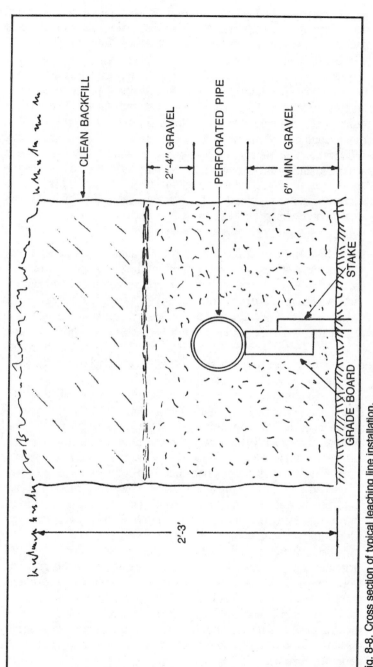

Fig. 8-8. Cross section of typical leaching line installation.

CLEAN BACKFILL

2"-4" GRAVEL

PERFORATED PIPE

6" MIN. GRAVEL

STAKE

GRADE BOARD

2'-3'

311

Frequently that marks the end of the piping, and the usual procedure of adding a few inches of gravel, an interface and the final backfill can be completed. In other instances, though, a tailpipe is added to join all of the end of the drainpipes together in an effort toward improved effluent flow and distribution as well as air circulation in the lines. If this is the case, the tailpipe is put together in an identical manner to the header, using tees and elbows. The connections of the drainpipes to the tailpipe need not be leakproof, but those along the tailpipe should be. Usually the tailpipe is made up of solid rather than perforated pipe sections, and no gravel is needed around it since it is not part of the leaching field. But to reverse the statement, there are indeed occasions where the tailpipe is made of perforated pipe and is part of the drainage system, in which case it is surrounded with gravel like the other perforated pipes.

Seepage Pit and Grid System Differences

About the only difference between a subsurface-tile grid-type system and a seepage pit is the depth of the pipe. Seepage pits are usually made in grid fashion, and the grids are constructed just the same way as they are in a subsurface-tile system. Instead of being in trenches, the piping grid is laid out on a large continuous bed of gravel that extends outward about 3 feet beyond the outermost pipelines. Grade boards are usually installed to maintain proper pitch, and in other respects the installation is completed in the same way as the others just discussed.

Deep beds are also made in grid fashion, usually fairly small in size and with the drainpipe lines only a couple of feet or so apart. A grid measuring approximately 10 or 12 feet square is probably about average. Frequently the depth of gravel placed beneath the pipe is far greater than in a subsurface-tile system, and may be 2 feet or more. Grade boards are not used because of the depth of the gravel below and also because the individual drainpipes are relatively short and stiff, and hold their positions well. Pitch is not especially critical, and the grid might lie dead level or pitch slightly downward anywhere up to about ⅛-inch per running foot.

The grid may be open-ended or closed with a tailpipe. A much larger quantity of gravel is frequently placed above the grid than would be used with a subsurface-tile system, perhaps 3 feet or even more. In essence, the whole affair becomes a dry well of sorts. An interface is laid in place to keep backfill dirt from filtering down into the gravel. The sewer outflow line leading into the grid is generally placed at a 45-degree angle or greater. A distribution box is not generally installed.

VENTING SYSTEM

In some areas a venting system is required to be installed as part of the leaching field drainpipe network. In some cases this may involve only one or two vents, while in others several are put in.

These vents are made from the same pipe as the rest of the drainpipe system, joined at appropriate points to the drainpipes with tee fittings and solid, leakproof joints. The stubs of pipe are set vertical and protrude above ground level by about 12 to 18 inches. The top of the pipe is capped with a special hood to prevent debris from dropping down into the drainage system (Fig. 8-9). The location of these vent pipes may be specified in the local plumbing code. They are sometimes placed at the ends of branch lines or grid legs, or may be inserted in a tailpipe line. From a practical standpoint they can be located at almost any point in the system where they will provide reasonable air flow.

The purpose of these vents is partly to exhaust the slight amount of gases that might be present, but more importantly to admit fresh air into the drainpipe system and induce better air circulation. This aids both in evaporation of the effluent and oxidation of the substances carried into the drainpipes.

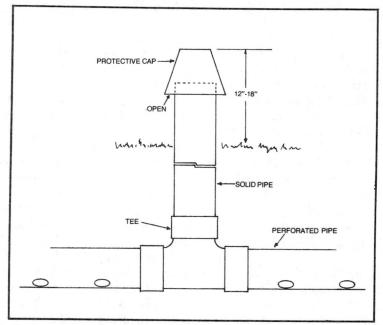

Fig. 8-9. Typical leaching line vent arrangement.

MUNICIPAL SEWER CONNECTION

Connecting to an existing municipal sewer system is more practical, less expensive and easier to do than building a whole septic system. However, for effective operation and a lack of problems, the job must be correctly done. In some areas the property owner can make his own installation, while in others the work can only be done by a licensed plumbing contractor or by employees of the municipality or the company that operates the sewage system and treatment plant. But whoever makes the installation, it is incumbent upon the property owner or his architect to take the initial steps in determining the sewer line layout.

Layout

When the plans for a new house are being started, one of the first considerations lies in establishing the house sewer line layout and its positioning relative to the sewer main to which connections will later be made. This layout may very well dictate not only the details of the house drain-waste-vent system, but in many cases can influence the positioning of the building itself.

The first step is to contact the local building department or city engineer's office to find out exactly where the closest sewer main lies and its exact depth. You should also find out if an outlet stub to service your particular piece of property has already been attached to the sewer main when it was laid. If so, you will need to know its exact location. This will be the starting point (or ending point) of your house sewer line. In the absence of such a stub, you can choose any convenient point along the main unless the authorities dictate otherwise.

Just as with a septic system, the house sewer line must be pitched downward from the house to the sewer main at a rate of ¼-inch to ⅜-inch per running foot. Thus, the point of connection of the house sewer line must be higher than the connection at the sewer main. Depending upon the topography involved and the differences in elevations between the main and the proposed building site, the relative position of the house drain, the height of the house foundation and/or first floor or perhaps even the location of the building itself may well be influenced. If there is any possible way to do so, the end of the house drain should be arranged to lie above the sewer main inlet, by a height equivalent to ¼-inch to ⅜-inch per running foot of distance between the end of the house drain and the sewer main inlet.

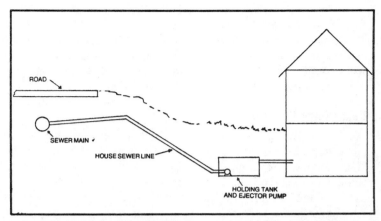

Fig. 8-10. Representation of municipal sewer line hookup where main house drain lies below sewer main.

If for some reason this cannot be done, there is an alternative. If the house drain must lie lower than the sewer main, the house drain is connected to a large buried soil tank that will continuously store a large quantity of raw sewage. The sewage is periodically discharged by an automatic soil pump, upward into a sewer line that rises above the level of the municipal sewer main and then pitches back down to join with it (Fig. 8-10). The cost of this additional equipment is sufficient to be reckoned with, and there are certain annual costs of operation and maintenance as well. And in the bargain, there is always the problem of what to do when the power fails. The soil pump stops working and the tank overfills.

To start the layout proceedings, first choose a likely looking route for the house sewer line to follow. A straight line is preferable, but bends can be introduced as necessary, keeping in mind that the straighter the line is, the better. Next you need to know the distance from the connection point at the sewer main to an appropriate spot at the proposed foundation wall or the termination point of the house drain. Usually the latter point is the easiest to work with, and the interior drainage system can be figured backwards from there. If the terrain between the two points is relatively flat, all you have to do is measure off the distance over the course of the house sewer line with a tape measure. If the ground slopes upward or downward, gaining or losing a substantial amount of elevation, you'll have to run a taut line between the two points, leveled with a line level. In many cases this operation is best done with a transit and rod, in the interest of both ease and accuracy.

Once you know the length of the pipeline run, you can easily determine how much higher the inlet end should be than the outlet end. If the run is 100 feet, for example, the difference will be 100/4 inches for a ¼-inch pitch or 300/8 inches for a ⅜-inch pitch. This works out to a 25-inch difference in the first case, and a 37 ½-inch difference in the second case.

Determining Location

Given the difference in elevation between one end of the pipeline and the other, you can now determine the relative location of the upper end of the pipeline, or the elevation of the joint between the house drain and the house sewer line. If the sewer main is located 6 feet beneath the surface of the ground, for example, the inlet of the house sewer line must be positioned at a point 8 feet 1 inch above a level line from the sewer main to a point beneath the building, assuming a ¼-inch pitch.

Obviously you can't draw a level line underground, but you can take a different approach to determine a specific point. Set a market of suitable height, say a 36-inch-high saw horse, over the sewer main location. Run a taut line back to the approximate location of the inlet end of the house sewer line. Make sure that the line is dead level. If the sewer main is 6 feet deep and the saw horse 3 feet high, the string is therefore 9 feet above the sewer line. The end of the house sewer line must lie 25 inches above that 9-foot level, or 6 feet 11 inches below the string. Consequently, the interior drainage system of the building, as well as the foundations and perhaps other structural details as well, will have to be adjusted to serve that elevation point (Fig. 8-11).

In cases where there is a substantial rise in elevation between the house site and the sewer main, and depending upon the length of the house of the house sewer line run, maintaining a continuous downward pitch of ¼-inch or so might not be possible. Depending upon the circumstances there are two possibilities, each each of which is the same as might be used in a septic system under similar circumstances. If the run is relatively short and the difference in elevation substantial enough, it might possible to pitch the entire sewer line at a continuous 45-degree angle or greater. If that can't be done, the alternative is to arrange the first portion of the house sewer line at a standard gentle pitch, such as ¼-inch per running foot. Drop the last several feet of pipeline down at a 45-degree angle with a reverse angle and short stub at the outlet end to connect to the sewer main (Fig. 8-12).

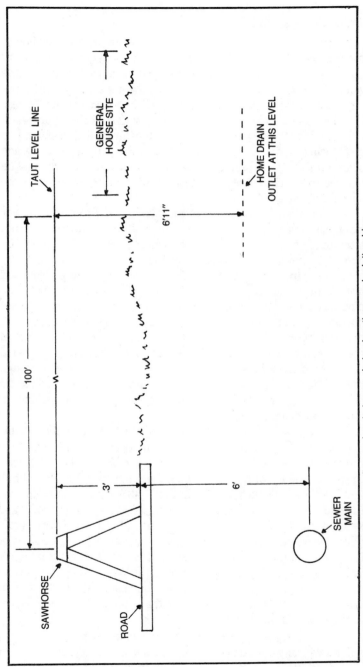

Fig. 8-11. Method of determining sewer line and main house drain elevations and relationships.

317

Sewer Line Installation

Installation of the house sewer line begins with excavation of the trench. If a street cut is involved in the excavation, be sure to check first with municipal authorities to determine the rules of the game. Find out whether or not you need a permit or whether the street cut perhaps might have to be made by municipal employees.

The trench itself should be carefully dug to the appropriate pitch, so that the sewer line can be laid as much as possible upon undisturbed native subsoil. Or, if you wish, you can set a grade board for the pipe to rest upon. In this instance, the trench must be made a few inches deeper than otherwise in order to place the pipeline at the correct level.

If a connection stub has already been installed on the sewer main, all well and good. If not, the do-it-yourselfer should not attempt to cut into the pipe; to make his connection; this should be done by municipal workmen. Once that is squared away, the first section of pipe can be connected to the stub and the pipeline laid back to the building. Joints must be made tight and leakproof. The pipe is connected to the house drain at some appropriate point outside the foundation, or it may continue into the building proper to make that connection. The sewer line can pass under the footing of the foundation and out through the floor, and in this case is installed before the foundations are made. The pipe may be brought through the foundation wall at a later date, through a sleeve provided for the purpose in a poured concrete wall, or by simply removing a block in a concrete block foundation.

If the pipe passes into the building at some point above footings, plenty of support must be given to the pipeline where it crosses the open space that will later be backfilled full of loose dirt. Prop the pipe up with a stack of concrete blocks set firmly, plumb and level, on undisturbed subsoil (Fig. 8-13). Without this support, the weight of the backfill, along with its steady compaction and settling out, will drive the pipeline downward and cause a break or a pulled apart fitting. As you lay these sections of pipe, be sure to keep track of the pitch to ascertain that it is correct and that there are no sags in the pipeline, especially at the joints. Keep the pipeline straight, too. After the entire line has been laid, it's worthwhile to make a quick check to make sure that nothing has gone askew during the course of the job.

With everything checked out, the backfilling process can begin. As with any buried pipeline, the job should be undertaken with a considerable amount of caution in order to avoid damage to the pipe.

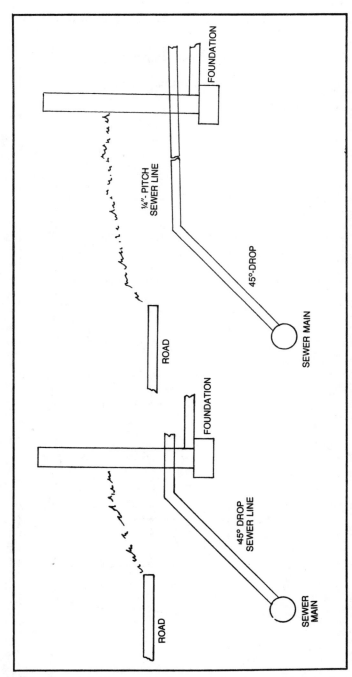

Fig. 8-12. When conditions permit, sewer line can be run directly from the house main drain at 45 degree angle (left). Where house elevation and sewer line length are too great to permit running sewer line at constant pitch, last portion of pipeline is a 45 degree drop pipe to sewer main.

319

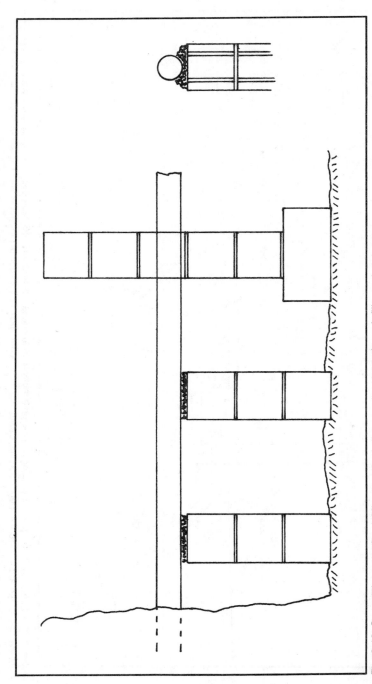

Fig. 8-13. Sewer pipe must be rigidly supported wherever it passes through fill area into house.

The first layer of fill should be clean and free of debris and rocks. If the sewer line is set on grade boards, hand-fill the trench to a depth of several inches over the top of the pipe. No compacting is necessary; and the remainder of the trench can be filled by machine. If there are large rocks present, though, don't dump them down into the trench. Ease them down gently, or push them aside and use them in the upper part of the fill. If the pipeline is set on undisturbed subsoil, hand-fill around the pipe and tamp the earth gently to help keep the pipe properly aligned. Then hand-fill with another foot of dirt, and finish the job by machine. Leave the last of the fill, preferably topsoil, humped up over the trench line, and eventually it will settle to approximate grade level.

Testing

Many a sewer line around the country is put in without benefit of any testing whatsoever, but in some localities tests are required prior to backfilling the trench. Specifics of these tests vary. If one is required in your system, you will be given all of the necessary details. Sometimes this involves merely filling the sewer line with water and checking for leaks. In other cases the test must be made with water under pressure of up to perhaps 10 or 12 pounds per square inch.

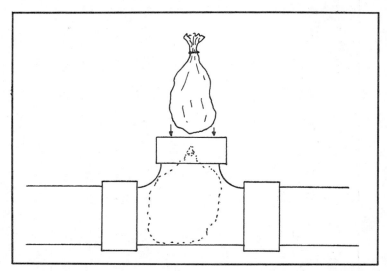

Fig. 8-14. Sandbag or other closure can be inserted into test tee located just ahead of sewer main connection to seal sewer line and DWV system for pressure tests.

Likewise, there are several ways to go about making such a test. One method is to install an open tee at the end of the house sewer line where it connects to the sewer main. The open leg of the tee faces upward. When the pipeline has been completed, the bottom end of the line is plugged by lowering a sack of sand or some other suitable closure device down into the tee opening, blocking off the sewer line (Fig. 8-14). The pipe is then filled with water and checked for leaks. Another method uses a special device made for the purpose, and the tee need not be installed. An empty air bladder is floated down the pipeline and into position at the end, tethered to a long length of flexible air line. The bladder is filled with air to close off the line, following which the test is made. Then the air is released from the bladder and it is withdrawn.

Whatever the specifics, you will probably be able to get some help from the sanitation department. Doubtless an inspector will have to be present while the test is going on. It's also possible that the department itself will conduct the test. In any event, don't forget and backfill the trench before required tests are made. You'll just have to dig the whole thing up again.

Chapter

Installing the Water System

Installing a residential water supply and distribution system using all, or nearly all, plastic components is probably one of the easiest jobs among the major home building projects that the do-it-yourselfer might accomplish. It just isn't all that difficult (though it is indeed somewhat time-consuming), especially once the system has been planned and laid out and everything is in readiness for the actual installation. But simple though it may be, there are a number of factors to keep in mind, as well as bits and pieces of general information, that will help in speeding the job to a successful conclusion. Though by this time you probably already have a pretty good idea of just how the system goes together and what you must do to make a complete installation, in this chapter we'll cover a few items having to do with the installation process.

SUPPLY LINE

Whether you begin your water system by installing the supply line or the distribution system within the house makes little or no difference; you can start with either subsystem. Of course, if your concern is only with replacing all or part of an existing distribution system, or with expanding that system, then the business of supply line installation will have little relevance for you.

The water supply line is installed in pretty much the same way for all water sources, but there are a few differing requirements that depend largely upon the type of water source being tapped. Also, the existence of local plumbing codes may dictate the inclusion of

...igs or the installation of the line according to certain
...s and/or specifications, that will vary with locale.

...ing To A Water Main

The first step in laying a water supply line that will be attached
...o a water main is to locate the main. This can be done with the aid of
municipal or water company authorities. If a tap stub already exists,
installed when the water main was laid in order to serve this
particular piece of property at some future date, you will probably
be required to use that stub whether it happens to be convenient for
your plans or not. If there is no stub, you can tap into the main at any
handy point.

The next step is to dig a trench, or have one dug, from the main
to the point of entry into the house. Obviously a great deal of caution
should be taken when digging with power equipment in the immediate vicinity of the water main itself, and also close to the house
foundation. Damage at either point is both frustrating and expensive. The trench bottom should lie below the local frost line, or to
whatever depth satisfies local code requirements. The trench floor
should be smooth and free of rocks and debris, and changes in
elevation should be made in gentle gradients, never in abrupt steps.

If a tap stub already exists at the water main, it will include a
stop valve, sometimes called a *corporation stop*. If no tap has been
made, a special saddle must be installed on the main with a stop
valve attached. The tap itself is made with a wet-tapping tool that
bores a hole in the water main without allowing water to escape. As
the tool is removed the stop valve is closed, and the job of laying the
water supply pipe can begin.

If rigid plastic pipe is used, start the pipeline with an expansion
loop of copper tubing. This attaches directly to the stop valve with a
flare fitting. The loop can be bent upward and back down to the
trench floor in the shape of a question mark lying on its side (Fig.
9-1). Attach a transition fitting to the end of the loop to join to the
plastic pipe. The same arrangement can be used with flexible plastic
pipe, though in some instances the plastic pipe itself can be buried
with a bit of slack to accomplish the same purpose. If the waterline is
buried beneath a roadway, the copper tubing is frequently extended
beneath the pavement before it connects to the plastic line.

Pipeline Placement

Now the pipeline itself can be laid. In rocky soil it is a good idea
to first lay about 4 to 6 inches of sand or fine gravel along the entire

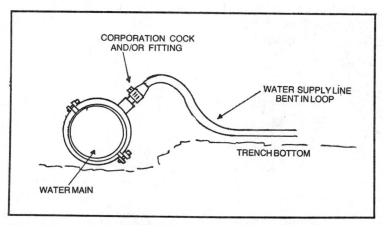

Fig. 9-1. Typical loop connection of water supply line to water main to allow expansion and contraction, as well as movement from ground or fill settling.

trench bottom. In fact, this is a good idea regardless of the soil type, though it is not often done. Rigid plastic pipe sections can then be laid out and welded together, taking care not to get sand or dirt into the fittings as the job progresses. If there appears to be much danger of this, temporarily elevate the pipeline of the trench floor with scraps of wood or whatever is handy. Double check each joint as you go along, to make sure that they are at least from outward appearances good joints. Flexible plastic pipes need only be unrolled from the coil along the trench floor. Allowing the pipe to snake back and florth slightly provides plenty of expansion-contraction room and will relieve any potential stress on the pipe. With either type, make sure that the entire pipeline is well supported from below, and does not rest upon jagged rocks or make any abrupt elevation changes.

At some point outside the foundation of the house the water supply line is usually attached to another stop valve, called a curb stop or curb cock. This valve, which is generally a brass one, merely lies on the trench floor with the valve handle pointing straight up, and is attached to the pipeline with a transition fitting assembly. From this point the pipeline continues through the house foundation. Here there may or may not be other stop valves, probably identical to the curb valve, but in this case called a meter valve.

Next in line is the water meter. If there is a meter stop valve, the water supply line will terminate there, with the water meter attached to the meter stop valve by a short coupling which can be either plastic or metal. If there is no meter stop valve, the water supply line will terminate at the meter. Another valve, this time a main shutoff valve of the gate type, is closely coupled to the water meter. If no meter is required, the water supply line will terminate at the meter. Another valve, this time a main shutoff valve of the gate type, is closely coupled to the water meter. If no meter is required, the water supply line terminates at the main shutoff valve. Connection will be made to any or all of these devices with transition fitting assemblies.

Checking for Leaks

The next step will let you know in a hurry whether or not you did a good job in laying the pipeline. Check first to make sure that all valves in the line are in the off position. The main shutoff valve handle should be fully turned down, and the "handles" of the stops (which only require a quarter-turn from full open to full closed) are at right angles to the pipeline. Start by slowly opening the corporation valve. You will hear a loud inrush of water which should cease almost immediately, depending upon the length of the pipeline to the next valve.

If you continue to hear a murmuring sound after the initial rush of water, look about and see if you can spot a geyser anywhere. If the initial sound stops shortly after you have cracked the valve open, open it all the way and then check all of the fittings and the connection to the next stop valve in line for leaks. If you see no problems, open the next valve and repeat the procedure until you reach the main shutoff valve in the house. Leave that one closed. Then wait a few hours and go back over the line again, checking for minor seepage. If you find any, turn off the water, drain the line and make repairs. Even a small amount of seepage can cause great difficulties later on, especially in a buried line.

Burying the Supply Line

Once you are satisfied that the water supply line is completely tight and leak-free, you can bury it. The first part of the job should be done by hand, regardless of whether the backfill material is free of rocks or not. The best bet is to initial cover. Failing that, pick and choose your backfill material carefully, making sure that it is free of rocks and debris, and cover the pipeline to about a 6-inch depth. The corporation stop is simply buried completely. The curb stop, however, receives a different treatment.

Before placing any backfill over the valve or the adjoining pipe, lower a cast-iron curb box down over the valve and seat it firmly.

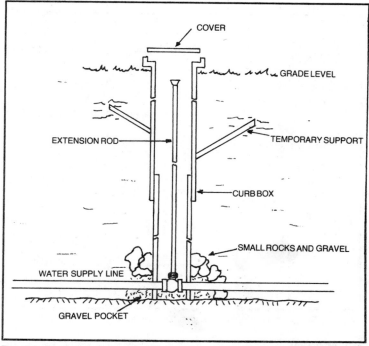

Fig. 9-2. Typical curb box installation; temporary supports are removed as trench is filled.

Prop it in place temporarily by the most expedient means. In some cases, an extension rod is attached to the valve stem and reaches to within a few inches of the top of the curb box, so that the valve can be operated without using a special, long valve key (Fig. 9-2). If this is the case, the rod should be attached to the valve stem prior to setting the curb box. Then the final backfill can be undertaken, either by hand or with the use of power machinery.

The first 2 feet or so of backfill material should be relatively free of rocks, and should be dumped carefully into the trench so as not to cause damage to the pipe. Rocky fill and boulders can be saved for the upper portion of the trench. Heap the material up 8 inches to a foot above grade level, and eventually the backfill will settle down to approximate grade level (Fig. 9-3). As you backfill around the curb box, remove temporary supports but hold the box in place so that it remains perfectly upright. Otherwise, the valve may be operable. As a last step, place the cover on the curb box. Be sure, too, to seal thoroughly around the pipe on both sides of the foundation when it passes through.

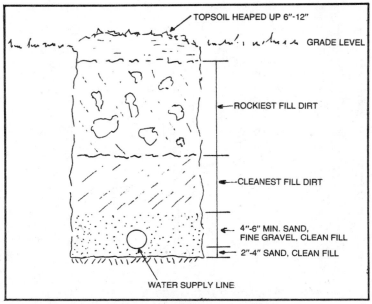

TOPSOIL HEAPED UP 6"-12"

GRADE LEVEL

ROCKIEST FILL DIRT

CLEANEST FILL DIRT

4"-6" MIN. SAND, FINE GRAVEL, CLEAN FILL

2"-4" SAND, CLEAN FILL

WATER SUPPLY LINE

Fig. 9-3. Cross section of water supply line installation.

Piping a Well

The basic installation procedures for a water supply line running from a well to a house are just the same as for a water main tap. The differences lie in the phyical setup of the well-associated gear and the presence or absence of certain fittings and equipment.

For instance, there are a number of alternatives as far as wells or similar water sources are concerned. A shallow dug well is often piped directly to the house, with the pump located in the basement. A drilled deep well, on the other hand, might be fitted with a submersible pump in the well casing itself, and be piped directly back from there. A well pit or a well house might be erected over the well, and the pump located with the well, in the pit, or in the residence (Fig. 9-4). There is no corporation stop, curb stop or water meter in such installations, but there may be one or more shutoff valves employed as a main valve or service valves. Unions may be provided if the pump lies in the line instead of at one end or the other.

The type of plastic pipe that is probably most commonly used for well installations is flexible polyethylene, and this can be used with the well and also for the underground supply line. Often the diameter is larger than that used for water main taps, and in most cases the inlet side of the pump requires a large-diameter pipe, while the outlet side requires a smaller size. Sometimes, too, the main supply line is run inside another, larger pipeline (1¼-inch slipped inside 2-inch, for instance). This affords extra protection for the water supply line. In case of damage or rupture from freezeup, the supply line can be pulled out of its casing and a new one slipped in without digging up the line.

To choose a random but common example of well piping, a shallow dug well can be pumped with a shallow well piston pump or a jet pump. A typical installation would go something like this. First, dig the trench from the well to the point of entry through the house foundation, taking care during the excavating not to disturb the wall of the well, particularly if it is of laid-up stone. Push a length of polyethylene pipe through a hole in the well wall at trench-bottom level, leaving a short stub protruding into the well. Make ready another section of pipe, calculated to the correct lenght so that the bottom of the foot valve and strainer assembly will be positioned a

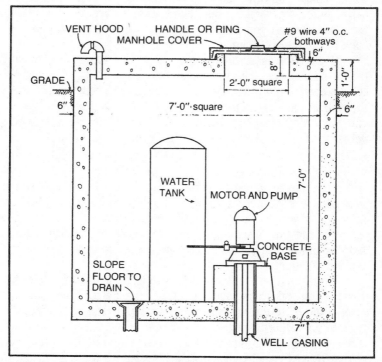

Fig. 9-4. Cross section of typical concrete pump pit and well pump installation (courtesy of Portland Cement Association).

foot or two above the bottom of the well, by attaching a foot valve with strainer at one end, and a 90-degree elbow at the other. Lower this assembly into the well and attach the free end of the 90-degree elbow to the supply pipe stub.

Lay the pipeline back to the house, through the foundation, and attach it to the pump. If desired, a stop valve and/or a check valve and/or a union fitting can be installed just ahead of the pump. Figure 9-5 shows a typical installation. Test and bury the pipeline in the same manner as described earlier. In this instance, however, the pump pressure tank must be fitted with a shut-off valve on the outlet side, so that the system can be closed in order to allow the pump to build up pressure and fill the tank. Refer to the pump and storage tank instructions for specific details.

COLD WATER SYSTEM

The starting point of the cold water distribution system is at the main shutoff valve, water meter or the output line from the water pump. You can begin the assembly of the system from this point, or you can install various segments of the system, connecting them together as you go along and eventually working your way back to the source point.

To take the assembly in logical sequence, the job begins with the main feed line, which most likely will be of CPVC plastic. Make the attachment to the source point with a transition fitting, and angle the pipe up and away on its course in the most practical manner. You can use a single riser as a main feed. Or you can run a feed line across the basement ceiling, for instance, and project upward with a series of risers, even to the extent of one for each plumbing fixture if you wish. You can also take any course you want to. Though you will often see an intricate series of neat right-angle corners in professional installations, they are totally unnecessary. Furthermore, a multitude of fittings cuts water pressure and adds to expense and time. Angle the pipes in whatever directions seem feasible and get the line to its destination in the most economical and practical manner.

The same holds true of the branch lines, and the feeders that travel through the living quarters of the house. Route them directly and wiggle them through structural cavities to the best effect. Be sure to give all the pipelines plenty of support in the process, and keep in mind the various problem areas that were discussed in an earlier chapter, such as support holes that are too tight-fitting to allow proper pipe expansion. Keep the number of fittings to a

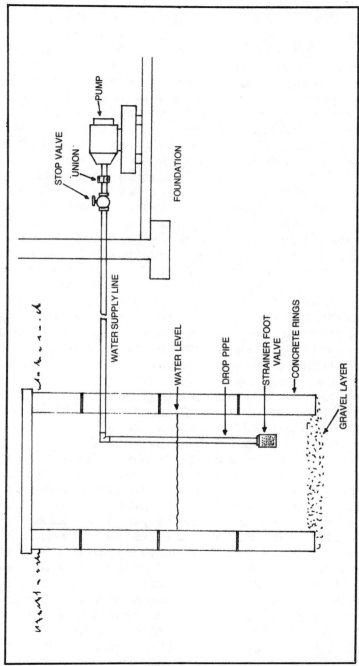

Fig. 9-5. Typical shallow or dug well water supply line and well pump installation with pump remote from well.

331

minimum, and use the optimum pipe sizes for the various feeds and branches that you settled upon during the earlier calculation and layout stages. Where pipelines pass through walls or other spots where trim moldings might later be nailed up, pictures hung or some other work be done, keep an eye peeled for the possibility of a nail later puncturing the pipeline. Either route the line above or below such points, or protect the pipe by attaching strips of metal to the edges of the framing members. Also, remember to stay clear of heat-producing sources, and wherever possible out of exterior walls in areas where rough winter weather is prevalent.

Use the dry-assembly approach, cutting, fitting and installing segments of the system dry before welding them up (Fig. 9-6). Once you have the risers and main feed lines in, with branch tees spotted at appropriate points, you can install each branch line from feeder to fixture stub-out. If the plumbing fixtures, or some of them, will not be installed and connected until later on, weld caps over the stubs to prevent the entrance of dirt or debris. You can then make a pressure test. In this instance it does no harm to make the stubs a good 6 inches long, which will give you plenty of pipe length to work with during fixture installation. An alternative to this method is to go ahead and attach stop valves directly to the stubs at the same time you install the branch lines, finishing up that segment of the job completely (Fig. 9-7). Stubs for those few fixtures that might not require stop valves can be capped.

Naturally, the location of each plumbing fixture should be carefully calculated so that you can bring the stubs out of the wall or up through the floor at the appropriate points. With regard to wall stubs, make sure that you know how much clearance you have behind and under the plumbing fixture. Give yourself plenty of room to install the risers later on. Don't inadvertently bring the stubs out of the wall smack in the middle of the backside of the fixture.

Once you have the entire cold water distribution system installed, make a last check to insure that you have not forgotten any segments, or left a fitting or two unwelded. Check to make sure that there are caps on all stubs, or that all stop valves are in the off position. Now comes the acid test. After allowing at least several hours and preferably a day to go by from the time you made the last welded joint, turn the water on at the main shutoff valve, or activate the water pump. Just as with the pressure test for the water supply line, you will first hear a strong inrush of water going through the pipes. If the distribution system is fairly extensive, this may last for several seconds but should slowly diminish.

Fig. 9-6. Wherever possible, it is a good idea to assemble piping segments dry. Then check for proper fit and alignment before welding the components together.

If the sound does not stop, you probably have a leak some-place. If the sound does not diminish at all, you have a wide-open line somewhere. Turn that valve or pump off in a hurry and investigate. If the sound does stop, the next chore is to inspect every section of the pipeline to make sure that you have no seeping or slowly dripping joints. If you find one or two, turn the water off, drain the lines, install a new fitting and try again. If there are no leaks, congratulate yourself and go on to the hot water distribution system.

HOT WATER SYSTEM

The hot water distribution system begins, logically enough, at the water heater. A sizable cold water feed line connects to the inlet side of either a tankless or tank-type water heater. The hot water feed line running to the remainder of the distribution system is connected to the outlet side. Since the tank-type heater installation is the most common, we'll use that as an illustration.

Setting up an electric water heater merely involves muscling it into its proper location and making the electrical connections at a nearby junction or circuit breaker box provided for the purpose, or wiring directly to the main entrance panel. Plenty of working room should be allowed around the area where the piping connections are to be made, usually at the top of the tank. There should be sufficient general working space around the tank as well. Though an electric heater can be placed directly against a wall, this is not recommended.

It is a good idea, too, to figure on room enough to add an insulating outer jacket to the tank after you have completed the installation. You can purchase ready-made fiber glass jackets at some hardware stores and plumbing supply dealers, or you can brew up your own with a roll of full-thick (3 ½-inch) roll fiber glass thermal building insulation. The object, of course, is to prevent as much heat loss as possible from the tank and thus reduce operating costs. This is particularly valuable where the tank is placed in a relatively cool location such as a basement. Setting up a gas-fired water heater is a bit more complex. Proper clearances from com-bustible surfaces must be maintained, a fuel supply line must be run in and connected, and a fresh-air vent and exhaust-gas flue must also be installed. Once these matters have been taken care of, the plumbing portion of the installation can be done.

Plastic pipe should never be directly connected to the hot water output side of a water heater. And though you might be able

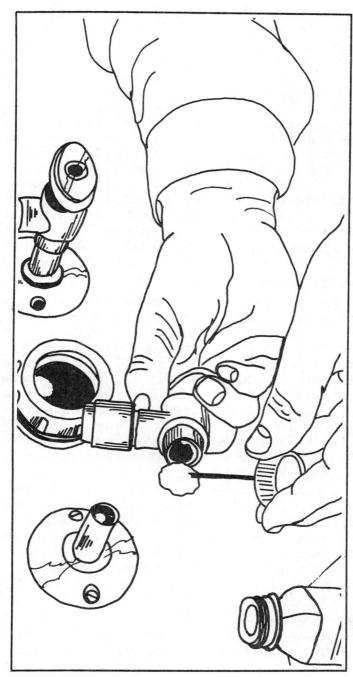

Fig. 9-7. Stop valves can be installed at the same time the supply lines are stubbed out, leaving only a simple fixture hookup to complete the job.

to get away with it on the cold water side, the usual procedure is to provide a metallic interface at both inlet and outlet. The inlet and outlet of the water tank will be equipped with standard female pipe threads. In some cases both inlet and outlet are at the top of the tank, while in others the hot water outlet is at the top and the cold water inlet at the bottom.

Whatever the case, use a galvanized steel nipple about 10 to 12 inches long along with galvanized pipe fittings to make quick directional changes if necessary (as when coming out of the tank bottom horizontally, and changing to a vertical pipe immediately beside the tank) to get clear of possible heat transference problems from the tank unit (Fig. 9-8). To the top of each pipe nipple, which can be purchased already threaded, attach a transition adapter, preferably of the type that will compensate for the different coefficients of expansion of the two dissimilar piping materials. Be sure to wrap each pipe thread with Teflon tape or use a liberal application of pipe dope.

Other Arrangements

There are some alternative arrangements that can be used here. For instance, if the cold water inlet is at the bottom of the tank, a short nipple can be installed with a tee on the end of the nipple. The cold water inlet is piped in the tee branch, and a drain cock is attached to the open end of the tee (Fig. 9-9). This allows a handy means of draining the tank. Also, it is a good practice to install a stop valve in the cold water feed line. This can be done with a brass valve at the end of the galvanized steel nipple, or with a plastic line stop inserted in the plastic portion of the cold water feed (Fig. 9-10). A stop valve can also be placed in the hot water output line. The valve should not be plastic, but rather a brass valve located between the galvanized nipple and the transition adapter.

In metallic piping systems, union fitings are also often installed in either the hot or cold water lines, or both. But many transition adapters will serve the same purpose as a union fitting, obviating the need for unions. If the transition adapters that you are using do not come apart into two separate threaded parts, consider installing conventional union fittings at the tops of the galvinized nipples so that the tank can be easily removed whenever necessary (Fig. 9-11).

Installing the Distribution System

Once work is completed on the tank installation, the process of piping the hot water distribution system can move along. This is no

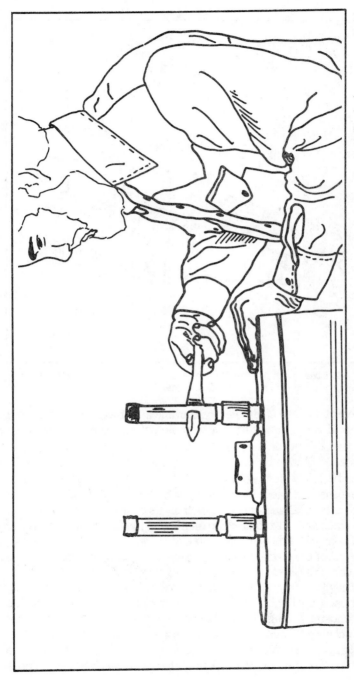

Fig. 9-8. Plastic pipe should not be directly connected to water heater. Start with long galvanized steel or other metallic nipples or connectors to allow heat dissipation.

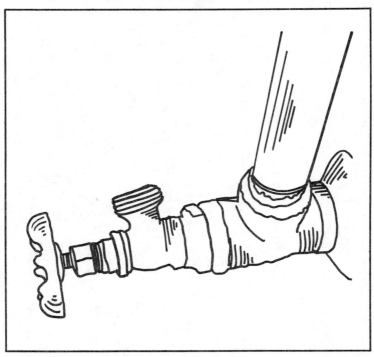

Fig. 9-9. Boiler drain or hose bib can be attached to tee at water heater cold water inlet to provide means for draining tank.

different than the cold water system and in fact will probably parallel the cold water lines to a considerable extent. Keep the two lines separated by 6 or 8 inches to prevent heat transference. Simply proceed section by section, fitting the parts together as you did with the cold water system. Pay particular attention to the inclusion of expansion doglegs as necessary in long pipe runs (remember that hot water pipelines should be as short as possible). Make sure that the pipes will not rub or chafe on framing members as they expand and contract; this will cause no end of chattering and squeaking.

After the distribution system has be completely installed, double check the lines to make sure that you have welded all the fittings and have not forgotten any stop valves or caps. Be sure everything is closed up tight. Go back to the water heater and check the thermostat setting; 110-120 degrees Fahrenheit is recommended for economy, but most dishwashers require on the order of 140 to 150 degrees Fahrenheit. Turn the cold water feed line stop valve to the "on" position. As the tank fills, check the tank inlet connections

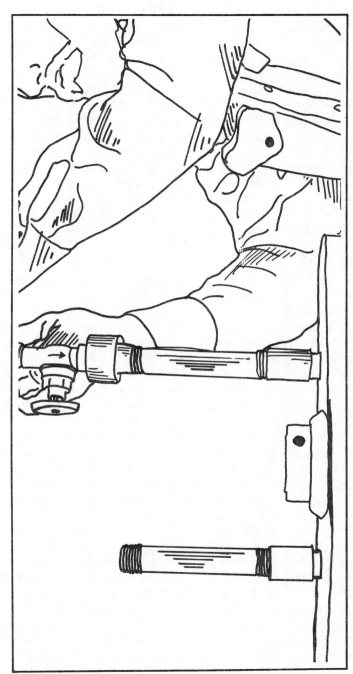

Fig. 9-10. Line stop valve should be placed in cold water feed line; another valve capable of 180-degrees Fahrenheit minimum temperature-handling characteristics can be inserted in hot water line if desired.

for leaks or seepage. If none appear, wait for the tank to fill and check the outlet connections. Eventually the lines will partly fill as well; check all of those fittings and stop valves for leaks or seepage. There will doubtless be air entrapped in the lines, and you can bleed it off by opening one or more of the stop valves. Have a bucket handy.

This is a good time, to, to make sure that the tank operates correctly. Turn on the electricity or fire up the bas burner, and let the tank run for awhile. Within a short while you should get at least a hint of hot water from one of the stop valves, and the galvanized steel outlet nipple should be getting warm to the touch. If everything seems to be properly operational, shut the tank off if you don't need the hot water at the moment. If an insulating jacket for the tank is part of your program, this can be slipped into place now. The job of installing a hot water distribution system is done.

PLUMBING FIXTURES

The actual installation of the plumbing fixtures is done in a variety of ways, depending not only upon the particular fixture in question but also upon the fixture style and manufacturer. There are countless minor variations, and the best procedure is always to follow the manufacturer's instructions to the letter, particularly observing all of the dimensional details. Many plumbing fixtures are not installed until surrounding finish work has been completed—a kitchen sink or a countertop lavatory, for instance. Bathtubs must often be installed during the construction process, since many tubs must be supported by the structural framework of the alcove into which they are built, and/or finish wall coverings must extend down over the outside lip of the tub rim. Shower stalls may also be built, at least in part, during the construction process. Toilets may or may not be left until all or most of the finish work in the immediate area is done. Water-using appliances are generally left until almost the last stages of finish work.

Likewise, the plumbing connections can be made at different times. Some must be made early on as a matter of convenience while frequently they can be left until last. By the same token, the water distribution and the DWV system can be installed and stubbed out before there is even a plumbing fixture in sight. Many of the fixtures can be installed before the plumbing system has even been started though in many instances the fixtures will then have to be moved and replaced in order to make the connections. So, as to the best arrangement for scheduling the plumbing fixture connections,

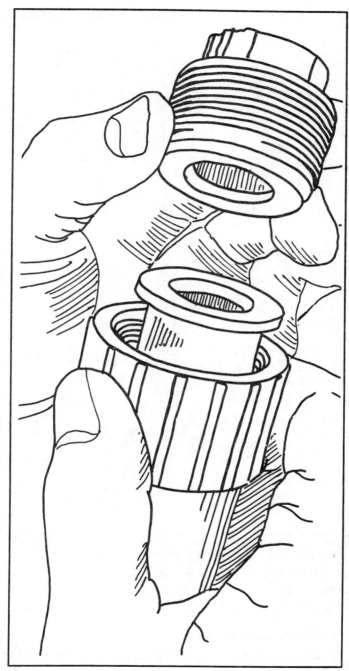

Fig. 9-11. Special transition adapters allow for different expansion-contraction characteristics when changing from metallic to plastic piping in hot water line, and also serve as unions.

341

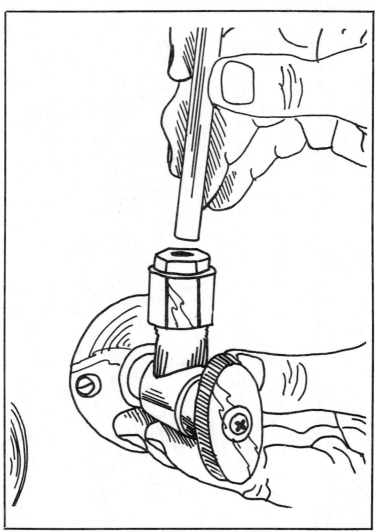

Fig. 9-12. Faucet connections can be made by simply inserting PB riser tube into stop valve fitting and snugging down the compression nut.

about the best that can be said is it all depends. Choose whatever sequence seems to be the best and easiest course for the particular fixtures that you plan to install.

Sink and Washbasin Connections

As far as the hot and cold water supply connections to the plumbing fixtures are concerned, there really is not much of a

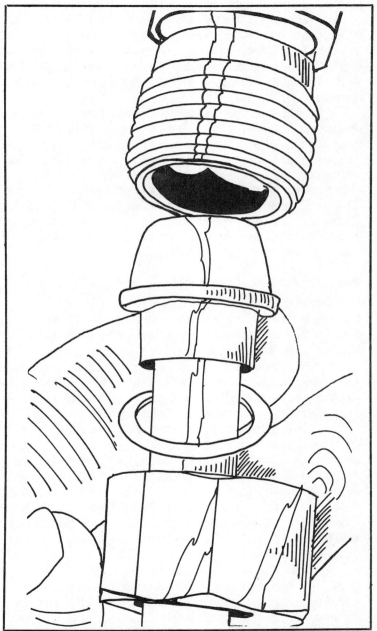

Fig. 9-13. Connection at fixture end of riser is a matter of slipping the specially formed upper end of the riser tube into the fixture supply tailpiece and tightening the compression nut.

problem. The easiest way to connect kitchen sinks and washbasins of whatever sort is to use polybutylene riser kits that are made just for this purpose. Sets are available in ¼-inch diameter (⅜-inch o.d.) in several lengths for direct attachment to plastic stop valves at one end, and standard faucet connections at the other (Figs. 9-12 and 9-13). The tubing is flexible, and very easy to install and can be trimmed to length if necessary.

If you dislike the appearance of polybutylene, you can purchase chrome-plated copper riser tubes that will also do the job. In that case, you must either make a transition from plastic to copper tube, or install stop valves that will accept copper tube at the outlet side and adapt to plastic pipe on the input side. Either way, installation consists of setting the riser and snugging up the fittings. The whole job is over with in a moment or two. About the only thing to watch out for here is to get the cold water supply connected to the cold water side of the faucet, and the hot to the hot. But even if you goof, changing them is not much of a project.

Bathtub and Shower Connections

Connecting the cold water supply to the toilet tank is done in exactly the same manner, using a single-tube riser kit made for the purpose. Connections are made in just the same way as for a sink or lavatory (Fig. 9-14). Making bathtub connections is a bit more complicated. There are a great many different kinds of faucet assemblies used with tubs, so the best bet is to proceed with the installation of the assembly according to the manufacturer's instructions. Basically, however, at some point on the assembly there will be two female standard pipe thread connections. Use the correct transition adapters to fit the plastic pipe directly to the faucet assembly. If a riser is to be included for a shower head, this can be done with either plastic pipe or metallic, and again is simply a matter of plugging in the proper fittings. Repeat the process in installing a shower generally ½-inch diameter, that threads into the faucet assembly or diverter valve at one end and joins the chrome-plated shower arm tube at the other, at whatever height is appropriate for your installation.

In the case of a stall-type shower, making the supply connections is not much different than for a bathtub, though the job may be a bit simpler. Again, there are various sorts of controls that might be installed, such as a pair of individual valves, a pair of valves joined in one body or a mixing valve arrangement. The shower unit itself may consist of a single shower head, or a stationary shower head com-

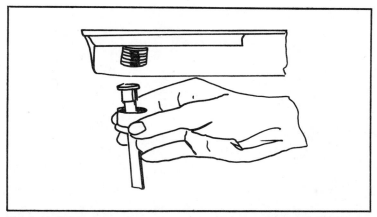

Fig. 9-14. Toilet tank connection uses single riser tube joined to tank with special fitting.

bined with a personal shower unit. But though specific minor details vary, connection basics are just the same. The hot and cold plastic water pipes must be adapted to the female pipe-threaded inputs of the faucets. A riser tube should be similarly connected to the valve assembly to serve the shower arm, just as with a combination tub/shower unit.

Appliance Connections

Connections to water-using appliances are also simple enough, although in many cases there isn't very much room to work in. Connecting a built-in dishwasher, for instance, can be exasperating if the stub-out isn't placed in just the right location. But again, making the connection is merely a matter of choosing the correct transition adapters to fit the appliance in question. Automatic washing machines are unquestionably the simplest to hookup since all they require is a pair of easily accessible boiler drains or hose bibs. Hooking them up is a matter of connecting the rubber hoses between the machine and the cocks.

In all cases where making appliance or plumbing fixture connections, it is an excellent idea to immediately make a pressure test of the joints you have just made up, especially if those joints will soon be hidden behind finish surfaces or trim plates. Turn the water on and leave the joints under pressure for at least several minutes. Check them for leaking or seepage. If all is well, your job is done. If not, best make repairs right now before you forget about it and the leaks cause some water damage.

Chapter 10

Installing the DWV System

Installing a complete residential drain-waste-vent system is a bit more difficult than installing water distribution systems, but not much. The pipes are larger and in some cases the runs may be more extensive, but the whole affair should go together with no problems at all. The key to any easy installation, particularly since the pipe sections are quite rigid and you are likely to be working in cramped quarters at least part of the time, is to make careful, accurate measurements so that all the pieces line up correctly and will fit together easily.

There are a number of possible starting points in assembling a DWV system, and which one you use makes little difference. By way of illustration, we'll assume that the house sewer line has already been brought to the foundation wall. We will use a typical simple system as an example. Actually, the more complicated systems for the most part are merely variations on or extensions of the same theme.

SOIL STACK

In such an installation, one might begin by accurately locating the position of the toilet. For ease of plumbing, the soil stack should be located directly behind the toilet, hidden in the wall. Of course, from a plumbing standpoint it could also just as easily be a foot or two to one side or the other, or in some other location entirely. Cut an accurately positioned hole through the bathroom floor of the proper size to accept whatever closet flange you plan to use. When locating this hole, be careful not to park it directly atop a floor joist, or in such a position that running the waste line from the toilet will be difficult or impossible because of other structural framing members. You need a clear, open space to work in here.

Carpentry Work

Next, spot the point where the soil stack will pass upward through the wall. Center a 3-inch (or 4-inch if you are using that size) pipe coupling or a short section of pipe over the point, and trace a circle around it (Fig. 10-1). Bore a hole a bit larger than the pipe diameter down through the wall sole plate and through the flooring into the open. The easiest way to accomplish this is with a speed bit chucked in a hand electric drill. Bore a series of small holes around the perimeter of the circle, and cut the scrap piece out with a keyhole saw (Figs. 10-2 and 10-3).

Now, dangle a plumb bob from the header plate of the wall so that the bob point is exactly centered in the hole you have just bored through the sole plate (Fig. 10-4). Mark the corresponding point on the top plate, and carve out another hole (Fig. 10-5). You may have to bore through a double plate here; be sure to check that you are not boring directly up into a joist lying atop the top plates. Wear goggles, too, because you'll get well sprinkled with chips.

In some houses this hole will bring you out into an open attic, but if there is a finished second floor, you will merely be in a ceiling cavity. If so, you must line up yet another hole to get through the second floor. Take a short section of 3-inch pipe with the end cut square and smooth, and apply a liberal dose of carpenter's chalk to the edges. Insert the pipe through the hole in the top plates, but don't touch the flooring above. Holding the pipe in one hand, set a level against the pipe and adjust the pipe until it is as vertical as you can get it. Then push it upward slowly, maintaining the vertical position, until it touches the floor.

Twist the pipe back and forth a few times. This will leave a chalked ring on the undersurface of the floor. Now you have a guideline, and can bore this hole out the same as you did the others. Since this is a difficult and awkward position in which to work, you might prefer an easier method. Locate the center of the chalked circle, and bore one small hole directly upwards. Make sure that the drill bit is vertical. Go up above, use the small bored hole as a reference point and draw a larger circle around it. Then bore the hole downward.

There is still one more hole to go. As you can see, installing a soil stack has as much to do with carpentry as it does with plumbing.

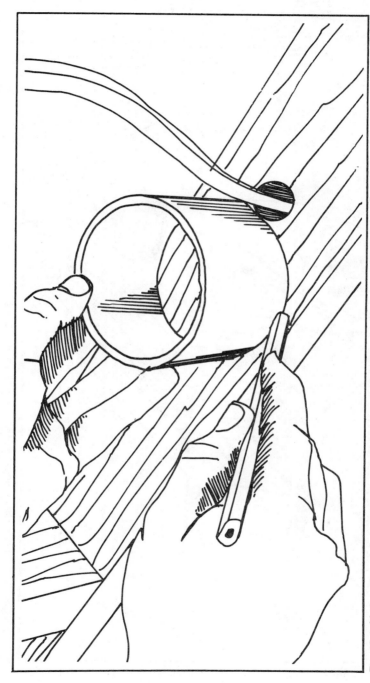

Fig. 10-1. Mark outline of drainpipe hole through sole plate by tracing around scrap of pipe or a coupling.

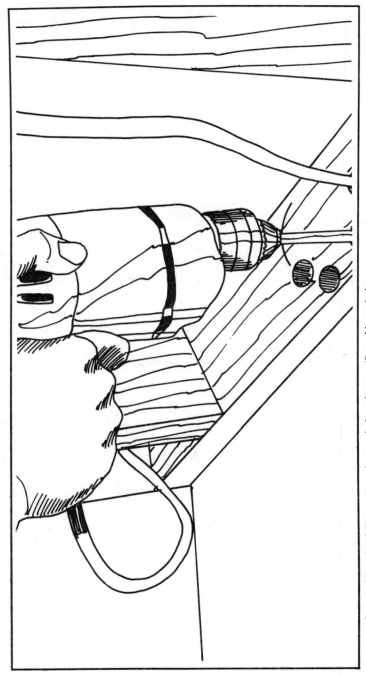

Fig. 10-2. Start hole cut by drilling series of small holes along outline of large hole.

With your plumb bob, locate a point on the underside of the roof sheathing that exactly lines up with the center of the hole you have bored through from below into the attic (Fig. 10-6). Again, using this point as a reference, trace a circle around a piece of pipe for a coupling to give you a guideline. Now, unless the roof is flat, this hole will not be large enough unless you keep the drill bit and the saw blade perfectly straight up and down as you work, a difficult and unlikely circumstance. The steeper the roof pitch, the harder it will be to get an exact fit around the pipe. The usual answer is not to bother with a tight fit, but arrange a loose one.

Bore a single small hole as nearly vertical as you can manage up through the roof sheathing. Then, gather up your tools and hie yourself up onto the roof. If the house is a new one and the finish roofing has not yet been applied, your task is an easy one. Hold a pipe coupling in a vertical position, centered on the hole you just drilled up through the roof. Trace around the coupling; this will leave an elliptical guideline on the roof sheathing or underlayment.

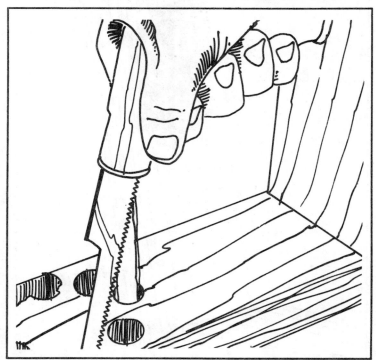

Fig. 10-3. Complete drainpipe hole by cutting through from drilled hole to drilled hole with keyhold saw.

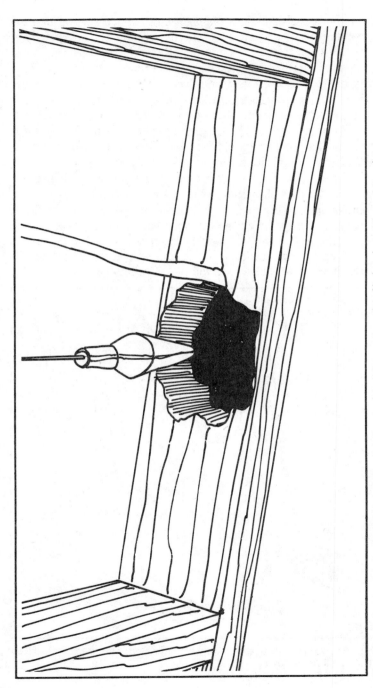

Fig. 10-4. Hang plumb bob centered in hole through floor to determine center point of hole to be cut through top plate.

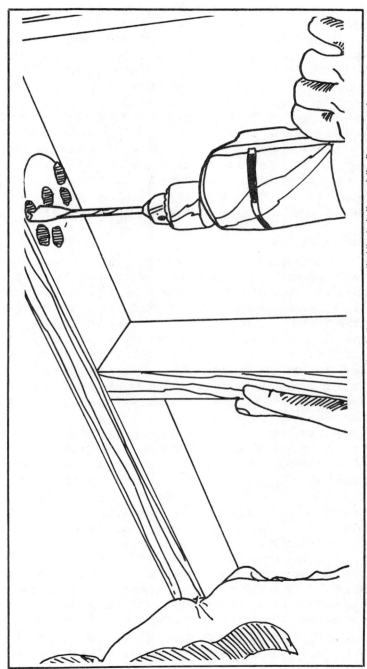

Fig. 10-5. Drill and cut hole through top plate and into attic or upper story in the same way that the hole through the floor was cut.

You can make the cutout with either a keyhole saw or a power jigsaw.

If you use the keyhole saw, you can saw approximately straight up and down, leaving the cut edges of the roof sheathing roughly vertical. The sheathing will be cut at an angle to its surface. Make your cut about ¼-inch to the outside of the guideline. If you use a power jigsaw, the shoe of the saw will lie flat on the sheathing and the cut edges will be at right angles to the sheathing, not on a vertical plane (Fig. 10-7). This means you will need more room to slip the pipe through, so make your cut about ¾-inch outside your guideline. If the finish roofing has already been applied, you will have to remove a section of roofing at least a bit larger than the dimensions of the bottom plate of the roof jack. Then you can go ahead and cut the hole using the same methods as just discussed.

Running the Pipe

Now it's time to start running the pipe. Here we run into any number of variables, depending upon your exact system layout, the specific kind and numbers of fittings required and so forth. But here's a typical approach for a simple system. Start at the sewer line stub, and run the main house drain pipeline back to the soil stack location. If the sewer line is the same size as the house drain, merely couple on the next section of pipe. Otherwise, make the joint with the proper adapter fitting. Drop a plumb bob down through the soil stack hole in the floor above, centered exactly. Take careful measurements, and end the run of main house drainpipe with a wye fitting. The slanted arm of the wye should be cocked upward, with a 45-degree elbow joined to it. The center of the opening in the top of the elbow should line up exactly with the point of the plumb bob. The lower, open continuation of the wye can be fitted with a cleanout plug, or the house drain may travel on from there to pick up another stack (Fig. 10-8). Remember that the exact location for this upturn in the pipeline is critical. Put the sections together dry and make a double-check to insure that you have them right before welding up the joints.

At some point just below the level of the joists of the first floor you will need another fitting, or perhaps a combination of fittings. There are any number of possibilities, such as a single or double special waste and vent fitting, a waste and vent tee or whatever. Calculate the location of the next fitting in line with respect to its proper alignment with the drainpipes that will eventually connect to it. Measure off the necessary length of pipe, insert the section into

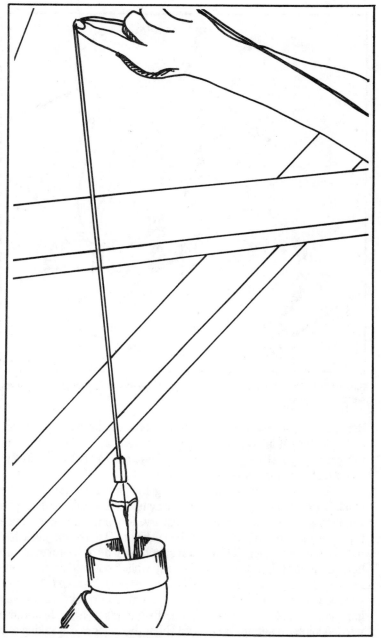

Fig. 10-6. Plumb bob centered in pipe opening or in hole through ceiling will determine center point of hole to be cut through roof.

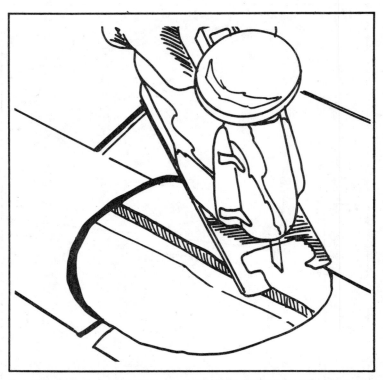

Fig. 10-7. Hole through roof can be most easily cut from topside with power saber or jig saw.

the upturned 45-degree elbow, and plant the required fitting on top. If the alignment is correct, weld them together. If other fittings in the same general neighborhood are required, install them in the same manner (Fig. 10-9).

Now you should be approaching the first-floor level. If the stack is to remain unbroken all the way up, the whole affair can be installed at once. Even in a single-story, low-posted house, more than one section of pipe will likely be required. One easy approach is to have someone else slide a 6-foot section of pipe down through the hole in the first floor and hold it steady while you weld up the joint. At this point, the end of the house drain line should be well-supported from beneath to prevent any undue strain from being imposed upon it. Remember, too, that it takes a while for the welded joints to set up. Be careful about the amount of yanking and tugging you do on a freshly-welded pipeline. The fittings can be jerked out of line all too easily in the early stages of curing.

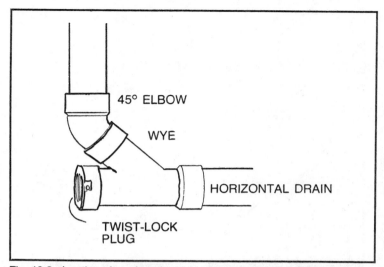

Fig. 10-8. Junction of stack and horizontal drain line can be made with wye fitting, which allows continuation of drain line or installation of cleanout. Extend stack upward with 45 degree elbow for easy drainage flow (courtesy of Genova, Inc.).

Once you have the first section of stack in place, weld a coupling to the top. Then repeat the process with another section of pipe. Have a helper slide a length of pipe down through the hole in the roof and give it some support, while you fit and weld it into the coupling. After the joints have cured for at least an hour, saw the pipe off 1 foot above the roof surface as measured from the highest point at the back or uphill side of the stack (Fig. 10-10).

Finish the job up by sliding a roof jack down over the pipe (Fig. 10-11). Snug the bottom of the jack tight against the roof underlayment. You can smear the bottom side of the upper two-thirds of the bottom plate with roofing compound if you wish. If you had previously removed the finish roofing, replace the material in such a way

Fig. 10-9. Install additional fittings and pipe sections as necessary below floor level.

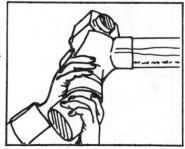

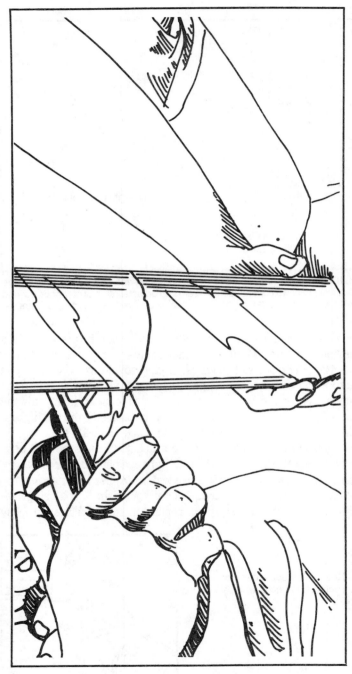

Fig. 10-10. The PVC-DWV stack can be lowered through wall and roof and anchored in place, then cut off with hand or power saw to 1-foot height afterward.

that the upper two-thirds of the bottom flashing plate is covered by the material, while the bottom third of the plate lies on the outside of the roof (Fig. 10-12).

In a two-story house, you will not get clear through to the roof with just one extra section of pipe; more will be needed. If that's the case, just keep adding sections by feeding them down from above and coupling the sections together.

Fig. 10-11. Roof jack fits tightly over pipe, seals out moisture.

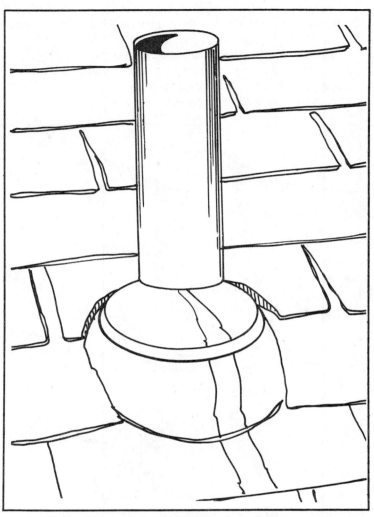

Fig. 10-12. Completed roof jack installation. Note placement of jack flange under upper shingles and over lower ones.

Often a soil stack does not run in an unbroken pipeline from basement to roof. In some areas it's permissible, for instance, to add a lavatory drain to a direct-vented system above the input level of the other fixtures on the line. This would require the insertion of a reducing sanitary tee or a reducing wye at some point a couple of feet above first-floor level. Other plumbing fixtures may drain into the soil stack from a second-floor level, or perhaps a vent tee must be installed at some point. Continuing the stack run in these circumstances is simple enough. Locate the proper position for the first fitting in the line, cut a suitable length of pipe, slip it down through the first floor and weld it into place in the uppermost soil stack fitting socket. Weld the fitting to the top of the pipe section and position it properly (Fig. 10-13). Repeat the process with the next length of pipe and the next fitting. Eventually, the last section of pipe will stick out through the roof and is treated in the same manner as just discussed with the full-length soil stack.

BRANCH DRAIN LINES

The next step is to take care of the branch drainage lines. This is done by working from the stack fittings back to the fixture locations. To start with the toilet waste pipe, you will need to join a sanitary elbow to a length of pipe which will fit into the proper 3-inch socket provided for it in the soil stack assembly. The length of the

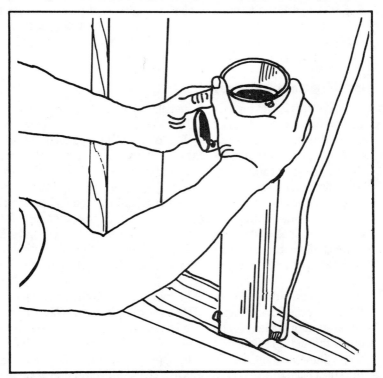

Fig. 10-13. Sections of stack and necessary intermediate fittings are positioned and welded, moving upward toward the roof.

pipe should be calculated so that the upturned opening of the sanitary elbow is exactly centered below the hole you earlier cut in the floor for the closet flange. Then a short stub of pipe must be welded into the upturned socket of the sanitary elbow. Its length should be calculated so that the closet flange can be slipped onto the pipe from above and welded into place at exactly the proper location and height, with a full-socket fit (Fig. 10-14).

Note that if the waste pipe lies just below floor level, a sanitary street elbow can be positioned to exactly match up with the closet flange without the extra short vertical pipe stub. In similar circumstances a street closet flange can be inserted into the socket of a sanitary elbow, provided that the upturned elbow socket can be fitted into the correct location. Once the joints have been welded up, the flange itself can be welded on and fastened to the floor (Figs. 10-15 and 10-16). Don't forget the drainage slope; even though the waste pipe may be quite short, that ¼-inch pitch must be present. Especially with the larger and shorter sections of pipe, this often means that the pipes are somewhat cocked in the fitting sockets. Don't worry about it. Just use plenty of solvent cement and everything will be fine.

Installing Small Drainpipes

The smaller drainpipes are installed in much the same manner. In a typical small and compact DWV system, there might be three such pipe runs left to make. These would be relatively short, consisting of a bathtub, tub/shower combination or stall shower, the bathroom lavatory and perhaps the kitchen sink in a back-to-back plumbing fixture arrangement. The kitchen sink and the bathroom lavatory are, generally speaking, almost identical arrangements. The usual procedure is to use 1½-inch diameter pipe for each drain line, even though the bathroom lavatory may have a 1¼-inch diameter drain flange. The drainpipe might stub out of the wall behind the fixtures, or come up through the floor below the fixture.

Whatever the case, locate the point at which the stub will exit the wall or the floor (be as accurate as you can in order to avoid later installation difficulties), and then lay out the pipe run. Bore holes down through the floor or the sole plate of the wall into the basement. Run the drainpipes from the appropriate fitting inlet socket at the stack back to an elbow with as few changes of direction along the way as possible. The elbow—a 90-degree fitting is fine here—will change the pipeline direction from horizontal to vertical. Slip

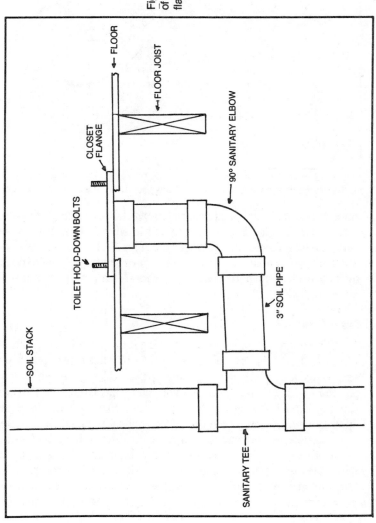

Fig. 10-14. Typical arrangement of toilet waste line and closet flange.

FLOOR

FLOOR JOIST

CLOSET FLANGE

TOILET HOLD-DOWN BOLTS

SOIL STACK

90° SANITARY ELBOW

3" SOIL PIPE

SANITARY TEE

363

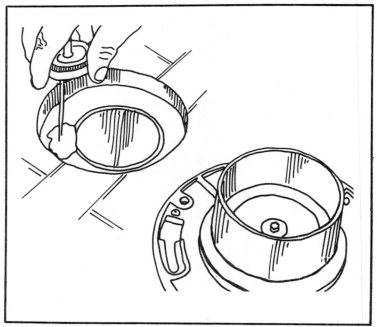

Fig. 10-15. End of toilet waste line is calculated to extend to just the right height so that closet flange can be welded on from above and rest flat upon the floor.

another length of pipe of appropriate length down through the hole in the floor or sole plate and make up the elbow joints. For a floor stub, that completes the drainpipe line (Fig. 10-17). In the case of a wall stub, you will need a second 90-degree sanitary elbow to make the turn back to the horizontal, plus a short length of pipe to stub out of the wall (Fig. 10-18).

Drain Line

The drain line for the one remaining item is handled in a similar manner. Bathtubs or tub/shower combinations are generally served by a 1½-inch diameter drainpipe, while a stall shower is plumbed with 2-inch pipe. Here you must calculate exactly where the trap for the shower or bathtub will terminate, and this depends upon the specific trap being used as well as its type. Sometimes drum traps are used with tubs or showers, and sometimes P-traps are used. The configurations of each are different. Once you determine the determination point and decide what fittings must be attached in order to turn the drainpipe line down through the floor, you can mark the appropriate point and bore a hole for the drainpipe.

Sometimes the trap connections are made above floor level, and sometimes below. Whatever the case, start at the one remaining stack drainage inlet and bring a pipeline of appropriate size back to the hole in the floor. Make a turn with a 90-degree sanitary elbow and stub upward to the connection point of the trap or make a direct trap connection as required (Fig. 10-19).

That completes the basic installation of the DWV pipeline for a simple system. In a more complex system, and depending upon local code requirements, there could be many more elements to be

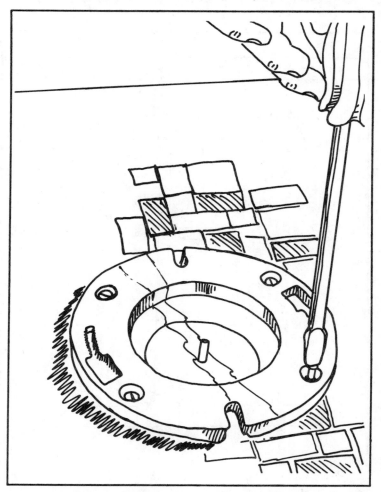

Fig. 10-16. After welding to toilet waste line, closet flange is secured to floor with screws.

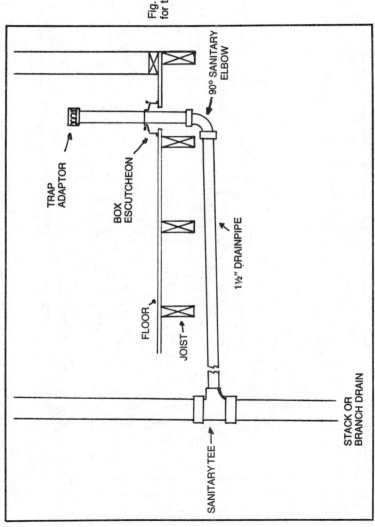

Fig. 10-17. Typical arrangement for through-floor sink drainpipe.

TRAP ADAPTOR

90° SANITARY ELBOW

BOX ESCUTCHEON

1½" DRAINPIPE

FLOOR

JOIST

STACK OR BRANCH DRAIN

SANITARY TEE

installed. There might well be additional plumbing fixtures, vent or revent lines, a continuation of the main house drain and the installation of another soil or vent stack, or the like. Whatever the specific case, though, the pipelines are installed along the same general lines as just explained. In all cases, the pipes must be well supported, the joints must be tight and leak-free, and a constant drainage pitch must be kept.

TESTING

The next question is, how do you know if the piping you have installed so far is really tight, and that there will be no leaks of either

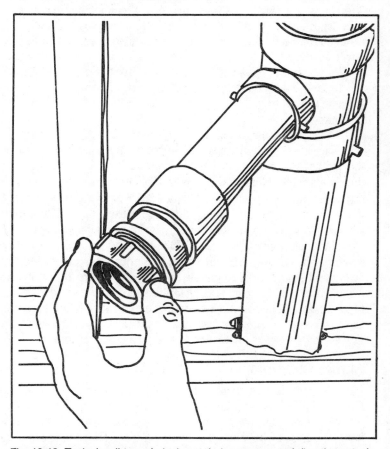

Fig. 10-18. Typical wall-type drainpipe stub, here connected directly to stack. Drainpipe might also turn downward with 90-degree elbow inside wall to make connection with branch drain beneath floor.

fluids or gases? Well, unless you make a test, you really don't know. In many areas, no testing procedure is required. In others, especially where plumbing codes are strict and rigorously enforced, a pressure test of one sort or another must be undertaken, observed by an inspector. He determines if the piping passes test (or can be repaired to the point where it can be certified). Even if not required, you might wish to make such a test.

If you plan to make a test of your DWV system, allow a couple of extra inches of drainpipe at every point where the pipes are stubbed out for later connections to plumbing fixture traps. Close off each of the open pipe stubs with a pipe cap welded solidly in place. Then the low end of the entire system must also be plugged. Often as not, a plug is put in at the point of connection to a sewer main. In the case of a septic tank installation, the plug can easily be put in at the tank inlet pipe. In some cases, especially where the sewer line has not yet been installed but there is a need to test the DWV system, the open end of the main house drain can be capped or plugged. The plug can be anything that will form a relatively watertight seal, such as a sandbag or an air bladder made just for that purpose.

Once all of the openings in the system are capped or plugged, in a typical test the next step is to insert a garden hose into the top of the highest soil or vent stack in the system. Let the water run until the entire system fills up with water. This will result in a relatively low order of water pressure on all of the lower joints in the system which, since the system is not under pressure in normal operation anyway, is sufficient to point up any leaks or seepage. When the system is full, start at the head of the main house drain (or at the sewer main connection) and check the entire system, fitting by fitting and segment by segment, until you are sure there are no leaks. Then pull the plug and drain the system. If there are leaks, repairs will, of course, have to be made. This simply underscores the attributes of doing a careful, methodical job of putting the system together, so that rework won't have to be done.

FIXTURE CONNECTIONS

The last step in completing the DWV system is to install the traps and make the final connections at the plumbing fixtures. The toilet is the only fixture that does not require an external trap, since it already has one built into the unit. To set the toilet, break out the plastic seal on the closet flange if there is one, put the toilet hold-down bolts in place, sweep the floor clean and unpack the

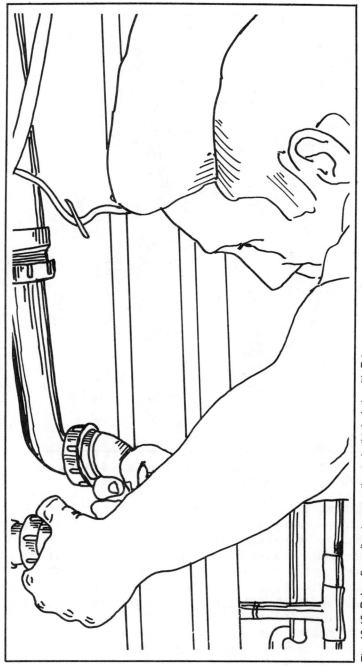

Fig. 10-19. Below floor direct connection to bathtub drain with P-trap.

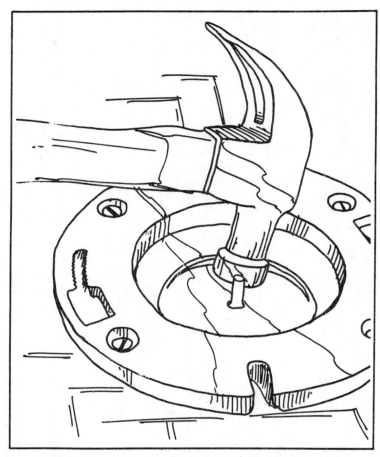

Fig. 10-20 First step in setting toilet is to break out closet flange seal. Not all flanges have seal, used for pressure testing.

fixture (Figs. 10-20 and 10-21). Place the toilet bowl, or the whole fixture in case of a one-piece unit, in place on the closet flange to make sure that you have a proper fit and all the right clearances for the fixture. Remove the toilet and set it on its side or top. Place a wax ring toilet seal on the discharge opening at the bottom of the fixture (Fig. 10-22). Center and square the wax ring so that it is accurately positioned, and press it firmly into place so that it will stick there.

Pick the toilet up and lower it gently over the hold-down studs and onto the closet flange opening (Fig. 10-23). Press the bowl down firmly but not too hard, and twist the bowl gently just once from side to side no more than a few degrees to set the wax seal. Then secure the nuts on the hold-down bolts. Be careful not to tighten them too much, as there is a danger of cracking the porcelain fixture. Just snug is enough. If the toilet is a two-piece unit, bolt the tank to the bowl, making sure that you get the rubber grommets and seals properly located. You may also have to anchor the tank to the wall. Consult the toilet manufacturer's instructions for details.

Lavatory and Kitchen Sinks

Lavatory and kitchen sinks are a bit simpler. Each must be fitted with an individual trap unit that attaches to the fixture drain flange at one end and the drainpipe stub-out at the other. If your pipe installation is perfectly accurate with respect to the fixture drain position, you can use a one-piece trap. However, a two-piece trap makes for easier installation because the parts can be swiveled in

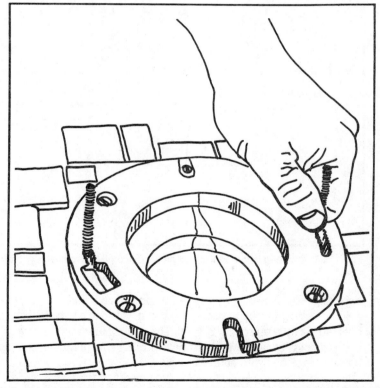

Fig. 10-21. Toilet hold-down bolts are positioned in closet flange slots and properly aligned to fit the particular toilet base.

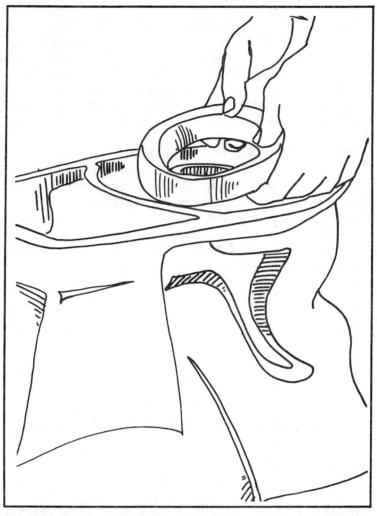

Fig. 10-22. Wax ring is positioned and pressed into place around toilet drain outlet.

opposing directions and can correct minor mismatches, or produce deliberate offsets. For a wall stub-out you will need a P-trap, while a floor stub-out requires an S-trap. There are numerous configurations readily available in hardware stores and plumbing supply houses, together with sufficient accessory pieces to enable you to make just about any kind of hookup you might need. Those made of polypropylene are the most effective and serviceable.

Making the actual connections is a simple matter. Start by determining the correct length for the stub-out, and trim the stub if necessary, in the process removing any caps that might have been installed. Solvent-weld a trap adapter to the stub-out. Then calculate the appropriate length for the tailpipe, the short length of drainpipe that extends down from the fixture drain flange. Trim if and as necessary, and slide a slip nut and a flexible compression ring from the trap assembly up over the tail piece. Fit the trap in place

Fig. 10-23. Toilet bowl is lowered carefully over hold-down bolts and light, slightly twisting pressure applied to seat wax ring.

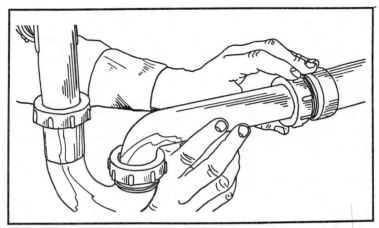

Fig. 10-24. Trap connections are made to fixture drain flange and drainpipe stub-out with trap adapter (on drainpipe); slip nuts and tailpiece of trap extension tubes as necessary.

and secure it loosely with the slip nut. Then check the length of the trap extension pipe, trim if necessary, and connect one end to the trap and the other to the stub-out trap adapter with the slip nuts and compression rings provided (Fig. 10-24). Adjust the trap assembly for the best fit and proper lie, and snug the slip nuts down hand-tight. Run a sufficient amount of water through the fixture to fill the trap and provide a seal, and at the same time check for leaks. If there is some seepage, simply tighten the slip nuts a bit further.

There are some slight variations in this system as far as kitchen sinks are concerned. Double-basin sinks can both be connected to one trap. This requires first attaching a piece called a continuous waste drainpipe to the sink. An end-opening continuous waste drainpipe drops directly down from one basin, while the second basin drainpipe angles into it. Each side is connected to the sink drain flange tailpipes by means of slip nuts and compression rings. The trap is then attached in the usual manner to the bottom of the vertical portion of the assembly (Fig. 10-25). A center-opening continuous waste assembly consists of a vertical centered drainpipe section opening into a pair of equal-length horizontal drainpipes, each of which attach to the basin tailpipes with slip nuts and compression rings (Fig. 10-26). The trap, again, is attached to the vertical portion of the assembly.

Various arrangements are also available for connecting the drainline of an automatic dishwasher into the drainpipe above the sink trap. Similarly, fittings are available to route discharge from a

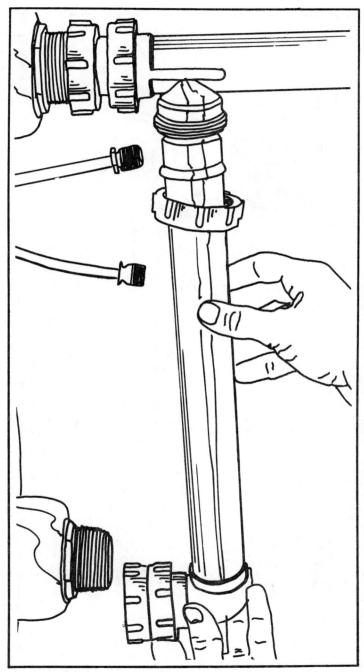

Fig. 10-25. End-opening continuous waste assembly allows single trap to be positioned to one side of the under-sink area.

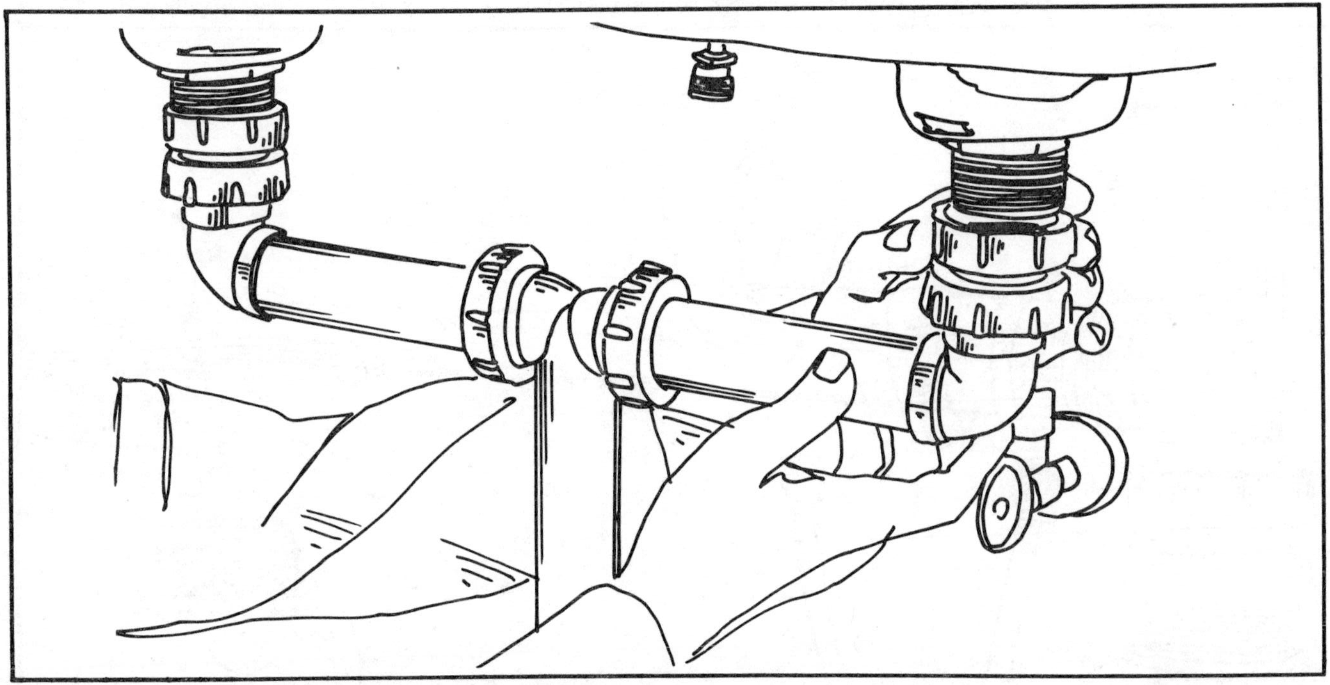

Fig. 10-26. Center-opening continuous waste assembly allows connection of two sink basins to one trap placed between the two basins in the under-sink area.

garbage disposal unit into the kitchen sink drain. Whatever the particular arrangement at a kitchen sink, there are various so-called tubular components available to make the necessary proper drainage connections.

Washing Machine

The drainage arrangement for an automatic washing machine is somewhat different than those for sinks or tubs. Usually the standpipe method is used, a simple but very effective arrangement. To arrange a washing machine drainpipe, first note the location of all openings, spaces or indentations in the back panel of the machine. This space can be used to house the drainpipe, so that the machine will cover the pipe and still set back against the wall.

After making the necessary calculations, bring a 1½-inch drainpipe line up through the floor in an appropriate position. Terminate the pipe somewhat above (or below) floor line with a trap

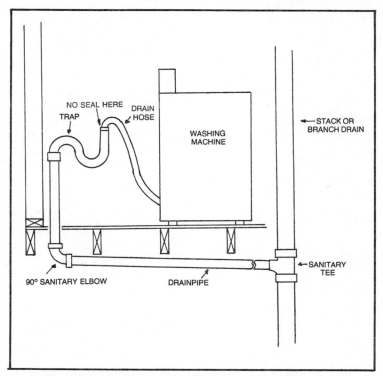

Fig. 10-27. Representation of washing machine drainage arrangement. In practice, trap would lie flat against wall and be hidden, or covered, by machine.

adapter welded to the pipe end. Attach a tubular S-trap to the adapter, and adjust the trap unit to the most convenient position. Then insert a short stub of drainpipe or a short tail piece into the top of the trap. Insert the crook of the washing machine drain hose into the top of this pipe stub (Fig. 10-27). There is no need to seal around the hose connection with putty or any other material; nor is this desirable. The trap itself provides a water seal. If the top opening is completely sealed up as well, no top venting will take place and the machine may not discharge properly.

Chapter 11

Miscellaneous Installations

We have covered all of those systems and subsystems of a residence that are commonly thought of as being, in combination, the plumbing system. However, there are other miscellaneous installations to be found in nearly every residence that also require plumbing materials and techniques. Some such installations are necessary, while others are merely desirable or take the form of additional do-it-yourself projects undertaken to increase comfort, convenience, livability or property value. None of these installations form an essential part of the basic plumbing system, but are instead auxiliary installations. Most have no direct connection to the basic residential plumbing systems, though there are a few exceptions.

GUTTERING

Not all houses are equipped with rain gutters. They were relatively common on older houses, but today are often installed on newer homes only as an afterthought. But many homeowners prefer to add rain guttering to thier houses, in some instances for the sake of appearance and/or convenience, or simply to upgrade the property. And in many cases, rain guttering is a necessity, or at least a doggone good idea. The guttering will divert rainwater runoff from the eave areas and well away from the foundations of the house, where the water can seep away harmlessly. This helps to avoid constant dampness of the foundation walls, along with seepage or leakage problems through foundation walls in areas of high rainfall. Guttering also does away with unsightly backsplash up onto the exterior siding, accompanied by drip gullies along the ground directly beneath the eaves.

The traditional material used in rain gutters and their accompanying downspouts is metal. Galvanized steel has long been popular. Copper has been used and aluminum with a baked-on paint finish has been in common use for some time. All of these materials, however, have their problems, most of which have been entirely overcome in the brand-new vinyl plastic guttering like Genova's Raingo system (Fig. 11-1). The gutter troughs, fittings, hangers and downspouts are all made of an off-white solid vinyl that will never rot, rust or corrode. They won't be harmed by ice in the system, are not prone to debris accumulation in the gutters, and are tough enough to withstand practically any ordinary bumps and bangs without damage. Though the color blends nicely with almost any exterior decor, the guttering system can be painted in any desirable color. Installation is literally a snap, since all of the parts do in fact snap together, except for the hangers and supports that are screwed to the building. The joints are made totally leakproof by means of special seals that are initially put together with a silicone lubricant.

Gutter Installation

Determining what you need to make a complete rain gutter installation is simple enough. Just measure off the length of each horizontal run of gutter that you will need. The gutter comes in 10-foot lengths, so judge the number of lengths you will need accordingly. Figure on installing one downspout for every 700 square feet of roof area. Downspouts also come in 10-foot sections. Measure the vertical drop from the roof facia, where the gutter will be mounted, to the ground. Plug in an allowance for the angled section that returns the downspout from the eave over-hang back against the wall, as well as the short discharge pipe at the bottom of the downspout. Multiply this figure by the number of downspouts that you will need, and divide by 10 to get the number of lengths of pipe required.

The gutters should be supported about every 2 to 2 ½ feet. You will need an appropriate number of hanger brackets, which attach either directly to the facia board or are suspended upon straps from the roof edge. Each joint between gutter sections needs a slip joint, and sections of downspout are joined with downspout couplers. In addition, you will need downspout brackets, a gutter drop outlet for each downspout, and an appropriate number of the

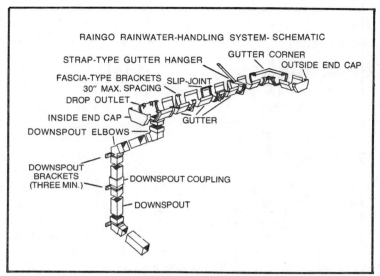

RAINGO RAINWATER-HANDLING SYSTEM- SCHEMATIC

STRAP-TYPE GUTTER HANGER

GUTTER CORNER

OUTSIDE END CAP

FASCIA-TYPE BRACKETS
30″ MAX. SPACING

SLIP-JOINT

DROP OUTLET

INSIDE END CAP

GUTTER

DOWNSPOUT ELBOWS

DOWNSPOUT
BRACKETS
(THREE MIN.)

DOWNSPOUT COUPLING

DOWNSPOUT

Fig. 11-1. All-vinyl snap-together guttering system like Genova's Raingo is long-lived, useful and easy to install (courtesy of Genova, Inc.).

proper end caps. Where the downspouts are angled, install downspout elbows. Two are required for each offset.

The actual installation is simple, though specific details vary a bit from job to job. Basically, the task consists of first snapping a chalk line along the facia so that the gutter brackets can be mounted in a straight line. The gutters should slope towards the downspouts at a pitch of about ¼-inch to the running foot. The gutter sections simply snap into place and are joined section by section with silicone-lubricated slip joints. The end caps are snapped in place, gutter drop outlets fitted, and the downspouts snapped together with whatever fittings are necessary. They are plugged into the gutter drops and secured to the exterior siding. Trim cutting of gutters or downspouts is easily done with a fine-toothed saw.

Routing Rainwater

Once the rainwater reaches the bottom of the downspout, there arises the question of what to do with it. The simplest method, of course, is to install a downspout elbow at the bottom of the downspout, attach a short tailpipe, and let the water drain out onto the ground. To prevent an unsightly gouge in the lawn, a cast concrete or flagstone splash pad can be set beneath the downspout to deflect the water flow (Fig. 11-2). Where drainage around

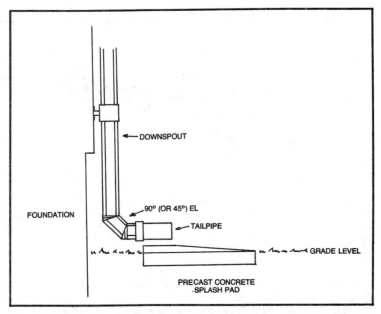

Fig. 11-2. Simplest method of rainwater disposal allows moisture to seep away into nearby ground.

the foundation is a problem, however, the best bet is to route the rainwater to a seepage point well away from the house.

Dig a shallow trench, starting at a point directly under the downspout, that travels out and away from the house for some distance in any convenient direction. The trench need only be a foot or so deep, and the trench bottom should slope down and away at a rate of about ¼-inch per running foot. Lay about 20 feet of solid plastic drainpipe in the bottom of the trench. Attach an elbow at the house end that faces directly upward beneath the downspout. Depending upon the guttering system that you are using, you may be able to purchase an adapter that will neatly make the transition from the bottom of the downspout into the plastic drainpipe. If not, continue the downspout section into the upturned drainpipe elbow about 3 inches or so (Fig. 11-3). Plug the gaps in the joints with a sealing compound if they are small, or with oakum and an elastomeric sealing compound if they are large. An alternative sealer is cement mortar or an epoxy cement. These materials do not allow for expansion and contraction of the materials, and are likely to crack open. The crack themselves, however, can be later sealed up with caulk.

At the other end of the plastic drainpipe, you have a couple of choices. One is to add a section or two of perforated drainpipe in a continuation of the trench. This section of pipe should be bedded in gravel, about 6 inches below and 4 inches above. Place a layer of roofing felt or straw (but not hay) on top of the gravel before backfilling the trench (Fig. 11-4). The interface layer will prevent dirt from settling down through the gravel and plugging up the pipeline. The arrangement will allow the rainwater to seep away into the ground and also to evaporate from the surface. The alternative is to construct a small dry well or seepage pit. The pit need not be very large, and can be done in a number of ways. The simplest

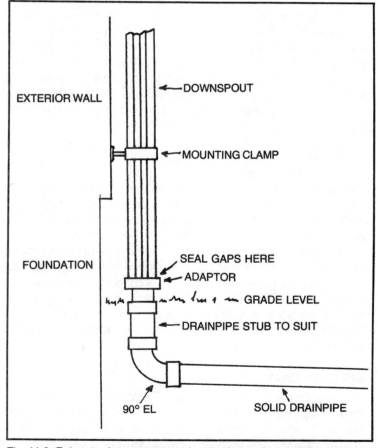

Fig. 11-3. Rainwater from gutters can be piped away from building in underground pipeline.

method is to dig a hole 2 or 3 feet in diameter and 3 or 4 feet deep, positioned so that the end of the solid drainpipe line lies about in the middle of the hole. Fill the hole with gravel up to about 4 inches above the pipeline. Put down an interface of tar paper or straw and backfill (Fig. 11-5).

Another possibility is to put a couple of feet of gravel into the hole and place an inverted old oil drum on top of the gravel. Cut a hole to admit the drainpipe at an appropriate point just below the top of the barrel. Poke or drill a series of ½-inch holes in the sides of the barrel. Push the barrel down into the gravel about a foot, and run the pipe into the hole. Fill around the drum with more gravel and backfill.

A third possibility is to dig a hole about 4 to 5 feet in diameter and about 4 feet deep. Place about a foot of gravel in the bottom of the hole, and line the sides with ordinary concrete blocks laid dry (no mortar) in a ½-overlap. Construct a simple lid of 2-inch-thick redwood planks that will rest on top of the top layer of blocks (Fig. 11-6). Set the cover in place and backfill all around. This type of seepage pit can be periodically dug up for cleaning without disturbing the pit arrangement at all.

SUMP PUMP

Many houses are afflicted with ground water moisture and drainage problems so severe that standing or running water in the basement may be a regular seasonal or even year-round problem. In such cases, frequently the only practical answer is the installation of a sump pump. This generally involves a considerable amount of labor, not to mention a bit of expense, but does mitigate the problem effectively. One of the best pumps to use for residential sump pump installations is a pump made almost entirely of vinyl plastic, such as Genova's Sump Witch (Fig. 11-7). The pump has the same excellent non-corroding and free-flowing characteristics of any plastic pipe. This type of pump is practically indestructible when properly installed and cared for.

Setting the Crock

The first step in making an installation is to determine where the maximum amount of water stands in the basement, and what its flow paths are. In some cases, it may be necessary to chip out concrete flooring and arrange recessed gutters to control the flow and direct it to one particular point, usually the lowest point, in the cellar floor. At this point a sump must be built. This is a matter of

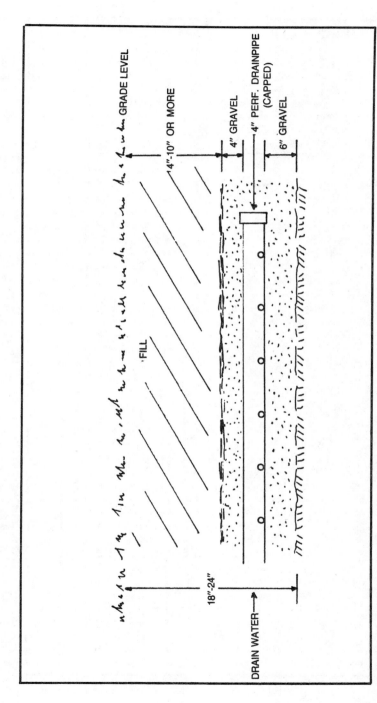

Fig. 11-4. Remote absorption trench can be made for rainwater disposal in similar fashion to septic tank leaching lines.

385

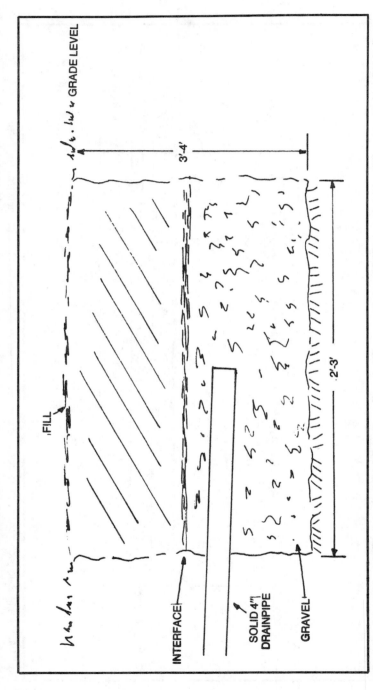

Fig. 11-5. Dry well installation for rainwater disposal.

GRADE LEVEL

3'-4'

2'-3'

FILL

INTERFACE

SOLID 4" DRAINPIPE

GRAVEL

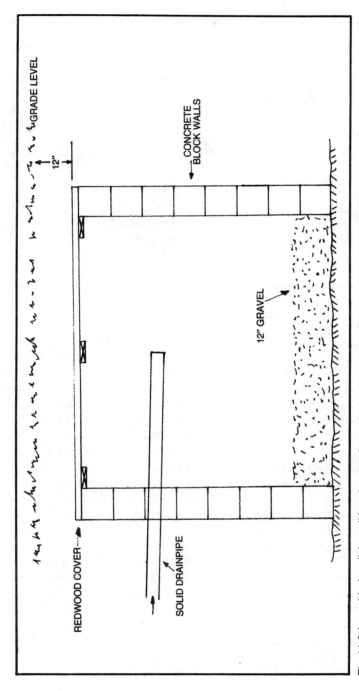

Fig. 11-6. Large block-wall dry well for rainwater disposal in large quantities. Block can be laid in rectangle and mortared together, or dry-stacked in circle.

387

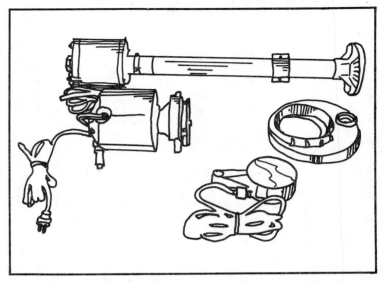

Fig. 11-7. Sump pump is the answer for keeping basements and foundations dry under severe moisture conditions.

chopping out the concrete floor and digging a hole in the ground beneath of a suitable size to accept a polyethelene sump crock, which looks a lot like a big bucket (Fig. 11-8).

The crock must be set in such a way that holes can be cut in the crock sides to line up with the drainage trenches in the cellar floor. If there are underground drainage lines, the crock must be able to receive those lines at the proper locations. The crock is set in the hole in the floor and properly aligned, with drainpipes routed into the crock as required (Figs. 11-9 and 11-10). Then a soupy mixture of cement mortar or concrete patching compound is poured in around the crock to stabilize it and add strength to the crock walls. Concrete patching compound is troweled around the top to fill in the ragged edges of concrete floor (Fig. 11-11). It is also smoothed around the floor gutters, if there are any.

Next the sump pump is lowered into the crock (Fig. 11-12). Cutouts are made in the crock cover for the pump standpipe and for the discharge pipe. The cover is slit and slipped into place. Electrical cords come through the same holes. A check valve is installed on the discharge outlet, and a suitable pipeline is attached to the check valve to carry drain water off and away (Fig. 11-13). Electrical connections are made by plugging into a grounded wall receptacle, positioned a short distance away on the wall.

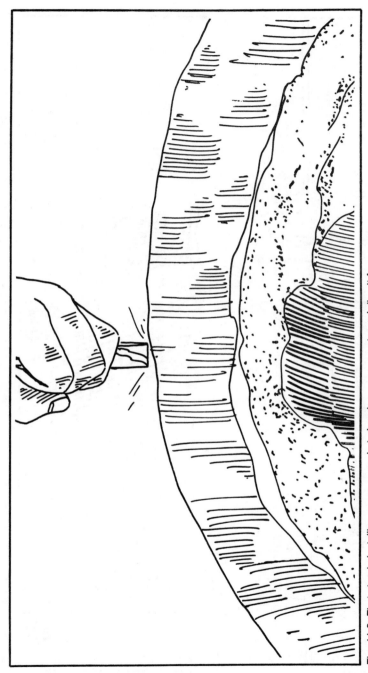

Fig. 11-8. First step in installing sump pump is to break away concrete and dig pit for sump.

389

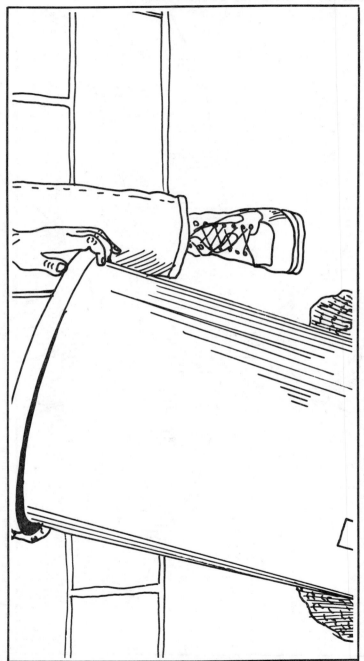

Fig. 11-9. Sump crock, which may be plastic or clay, being lowered into position.

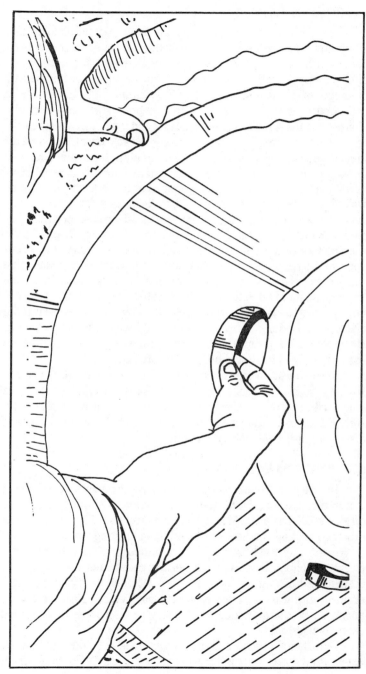

Fig. 11-10. Sump crock must be fitted and aligned to underground drainage input lines, as well as surface gutters if there are any.

Handling Pump Discharge

The drainage discharge from the pump can be handled in a number of ways. Where the house main drain connects to a municipal sewer line, the most expedient method is to tap the discharge pipe directly into the main house drain or to any other nearby drainpipe of suitable size in the DWV system (Fig. 11-14). If the main house drain is connected to a septic tank, however, discharge into the sewer system is not a good idea. This is especially true where the sump pump can discharge sizable quantities of water in relatively short periods of time. This simply puts too much of a load on the septic tank and is likely to cause tank operational problems.

Instead, route the discharge pipeline through the foundation wall at some likely point about 18 to 24 inches (or more) below ground level. Adapt the discharge pipe to 4-inch solid plastic drainpipe, and continue the line in a suitable trench pitched ¼-inch to the running foot down and away from the house. At some convenient point in the pipeline you can shift over to perforated drainage pipe laid in a gravel bed. The length of the perforated section of the drain line is dependent upon the volume of water that the sump pump can be expected to discharge. If the volume is low and particularly if the discharge is periodic, perhaps a section or two of pipe will do the job. However, if the volume is large and on a more or less continuous basis, the pipeline should be considerably longer.

The absorption capability of the soil also plays a part, of course. In many cases, especially where there is insufficient room to run a long perforated drainage line, a better solution may be to build a dry well. Either installation is put together in much the same way as was discussed in the section on guttering.

FOUNDATION DRAINS

In order to prevent moisture seepage or downright leakage of water through foundation walls, foundation or footing drains are often installed as a building is being constructed. The same type of drainage system can be retro-fitted to an existing building in order to correct moisture problems, too, although at considerable expense and effort. Sometimes, in serious cases, there is little choice if structural damage is to be avoided. Though heavy, fragile clay drainage tile was most often used in past years, plastic drainage piping has now largely superseded clay as a much more practical and efficient material.

Installation of such a drainage system is not a difficult matter, especially if done as a building is being constructed. The total cost of

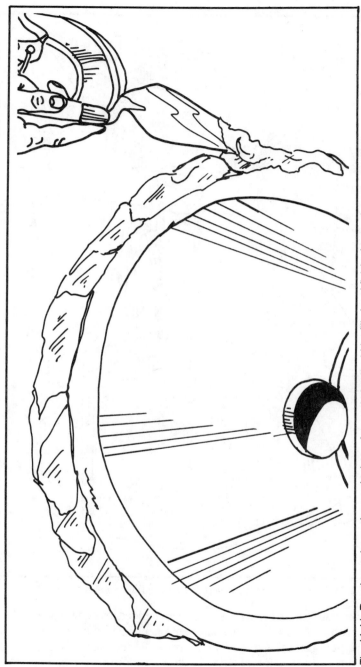

Fig. 11-11. Rough opening and ragged edges of sump hole should be filled and finished around crock edge.

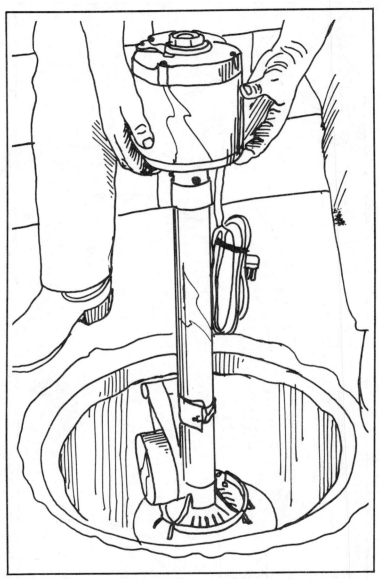

Fig. 11-12. Sump pump assembly being lowered into place in crock.

the system is so low that it is a recommended procedure in any new construction where there is even a faint indication that moisture might somehow eventually penetrate the foundation walls, either from the presence of periodic ground moisture or simply from potential runoff of rainwater from the building roofs. Rigid, perforated ABS or PVC drainpipe can be used for the purpose. From the standpoint of ease of installation, as well as a couple of technical points, corrugated, slotted polyethylene is even better (Fig. 11-15). The rigid types require the use of fittings to make joints

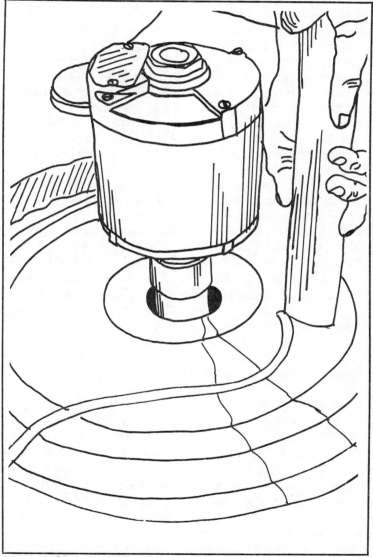

Fig. 11-13. Crock lid is fitted around pump head and discharge line.

between sections and to make directional changes. Flexible corrugated pipe can often be laid, even around corners, in continuous lengths. Either 3-inch or 4-inch diameter can be used; the latter is perhaps more common in residential applications.

The installation begins by preparing a flat surface about a foot or so wide, level with the bottom of the foundation footing or wall, and following it all the way around. A trench must also be dug for the drainpipe that will lead accumulated water away from the building to a disposal field or sewer line. The flat area next to the foundation should be covered to a depth of about 2 inches with gravel ranging in size anywhere from ½-inch to 2 inches or so, with a minimum of small particles. Then the perforated pipe can be laid in a continuum around the foundation. It is important that the entire run of pipe be pitched downward toward the collector pipe that leads to the disposal area. The minimum recommended pitch is about 0.2 percent, or 2 feet of drop in every 100 feet of pipeline. Thus, you can stay with the familiar ¼-inch per foot pitch so commonly found in other plumbing installations.

Once the pipeline is laid and properly graded, the next step is to cover it with more gravel. This layer should extend to the outside of the pipe for several inches, and completely cover it to a depth of about 2 inches. This allows a substantial gravel sump or collection area for the water, which is carried away by the piping as fast as it accumulates. Figure 11-16 shows a typical installation. Backfilling should be done with care, so as to avoid disturbing or perhaps damaging the pipes. Use only clean fill, and save rocky soil for the top part of the area to be filled. The collector pipe leading to the disposal field can be bedded in gravel or sand as a protective measure if that seems desirable, but it need not be buried in gravel. The pipeline itself must be solid, not perforated, and if sectional should be made up with tight joints.

Interceptor Drains

An interceptor drain serves primarily to intercept and divert either surface drain water or underground moisture. The reason for installing such a drainage system is simply to protect something from moisture damage. Usually the system serves the foundations of a building. It can also be employed to divert water from a tennis court area, lawns and gardens, outbuildings, a swimming pool or anything else that might stand in need of such protection. Whatever the specifics, the drainage system is placed upslope and roughly parallel with whatever is to be protected. It must be a sufficient

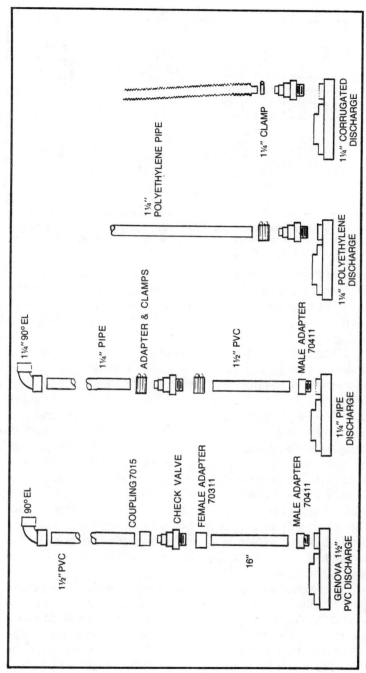

Fig. 11-14. Several methods of piping sump pump discharge line (courtesy of Genova, Inc.).

397

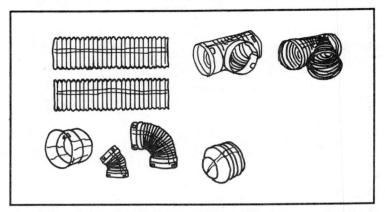

Fig. 11-15. Corrugated plastic drainage tubing and fittings. Top left, slotted and solid tubing. Right, snap tee and 45-degree "Y". Bottom left to right: snap coupling, elbows, split end cap.

distance away to provided good diversion in an efficient and unobtrusive fashion, yet close enough to collect the greater part of the expectant moisture flow. In most cases, the drainwater is channeled into a collector pipe that is routed to a convenient absorption or disposal area.

There are several ways to arrange an interceptor system, two of which are particularly effective and are also easy to build. In cases where it is desirable to divert moisture flow that might spill down (or through) a sloping area (in order to prevent muddy or soggy grounds around a house, for instance), a simple branching drainpipe field can be installed (Fig. 11-17). One or several lateral lines of drainpipe are buried about a foot below the surface of the ground, arranged so that they cross the slope and are set at a slight downward pitch, and are roughly parallel (if possible) to the area being protected. The downside end of each lateral curves into a solid collector pipe, which carries the drainwater off to a disposal site. The perforated pipes, which need be only 3 or 4 inches in diameter, should be fully bedded in gravel containing a minimum of fine particles; ½-inch size and up is the best bet. The open ends of the laterals should be capped.

If large amounts of surface runoff are expectant, this type of system probably will not do the job. Instead, a system called a curtain drain should be installed (Fig. 11-18). Here, a single trench is all that is necessary in all but extreme cases. The trench should be dug 24 to 48 inches deep and 18 to 24 inches wide, across the slope above the area to be protected. The locations should be as close to

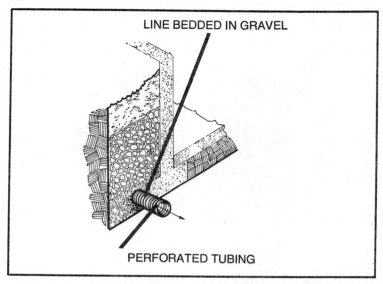

Fig. 11-16. Typical foundation drain installation (courtesy of Advanced Drainage Systems, Inc.).

the bottom of the slope as is practicable. Rather than being in a straight line, the trench should run roughly at right angles to the contour lines of the slope. The trench bottom should pitch downward toward the disposal area or the starting point of a collector pipe at a rate of about ¼-inch to the running foot.

The bottom of the trench should be smooth, and is best grooved or ploughed along its longitudinal centerline so that the drainpipe line can be nested down into it to about one-third of its diameter. The soil in the trench bottom should also be relatively

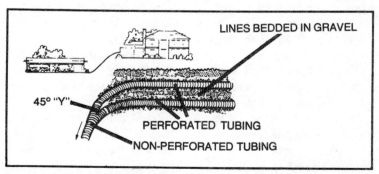

Fig. 11-17. Typical interceptor drain installation (courtesy of Advanced Drainage Systems, Inc.).

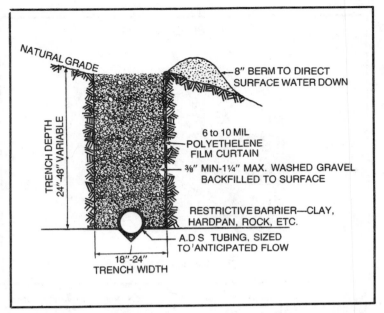

Fig. 11-18. Cross section of curtain drain (courtesy of **Advanced Drainage Systems**, Inc.).

impermeable. It should have low absorption capabilities like clay or hardpan. The downslope side of the trench should be covered with a layer—a curtain—of heavy-duty polyethylene plastic film. The black variety of at least 6-mil thickness and preferably heavier is the best bet. If the trench bottom is absorptive soil, the film can be continued across to the upslope side. During the construction process, the film is simply laid over onto the ground at the top of the downslope edge of the trench, extending beyond for about 12 to 18 inches and held down with rocks or a few shovelfuls of dirt.

The next step is to lay the pipeline. Then the entire trench should be filled to the brim with washed gravel ranging from a minimum of ⅜-inch to a maximum of 1¼-inch size. Compaction is not necessary; slight settling of the material does no harm. If considerable settling takes place, more gravel can be added to bring the level up. At least a portion of the spoil dirt that was removed during trench excavation should be heaped up in a continuous berm along the downslope of the filled trench. This dirt will act as a "backstop" for downflowing water that will direct all of the moisture into the trench to be carried off. Extra spoil dirt can either be graded off downslope, used for recontouring or trucked away.

Under-Slab Drains

Under-slab drains are an important part of the construction process wherever excessive moisture conditions are likely to be encountered in the soil lying beneath concrete slabs. Standing moisture, whether continuous or periodically recurrent, will keep a slab damp, lead to mildew and result in other unsavory interior problems. If there is a potential for freezing, structural damage may also occur. This type of drainage system can be employed beneath basement floors, under driveways and walks, and in conjunction with any other type of concrete slab, such as a garage floor or a patio. Perimeter drainage systems can be used in addition or not as conditions demand.

Under-slab drainage systems are generally laid out in grid fashion, with the grid somewhat smaller in size than the slab above it (Fig. 11-19). The perimeter piping should lie about a foot from the outside edges of the slab, with the crossing lines spaced in accordance with the pipe size and the amount of moisture that must be carried away. The corners are formed with 90-degree elbows, and the cross lines are connected with standard or bullnose tees. All of

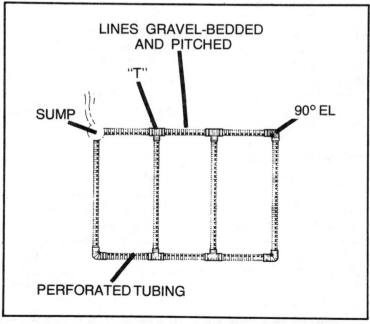

Fig. 11-19. Underslab drainage system layout (courtesy of Advanced Drainage Systems, Inc.).

401

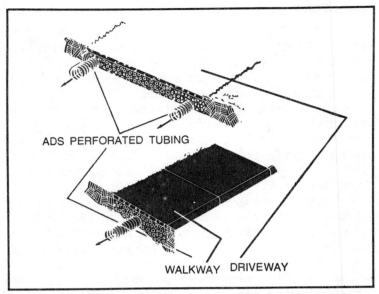

Fig. 11-20. Underslab drainage for walks and drives (courtesy of Advanced Drainage Systems, Inc.).

the piping is perforated. The entire grid should slope at about ¼-inch to the running foot at one or more collection points. Collection can take place through a solid pipeline that diverts the drainage water to an absorption field or other means of disposal, or in a sump in some convenient location (as in a basement) that can be equipped with a sump pump. Walkway and driveway drainage layouts are shown in Fig. 11-20.

RETAINING WALL DRAINS

A retaining wall is built to hold back the earth in a slope or banking. Because of its position, the wall is also likely to intercept quantities of both surface and ground moisture. Substantial amounts of moisture that collect against, under and in the immediate vicinity of a retaining wall can eventually cause considerable damage— either by undermining, deterioration of masonry joints or the unit themselves, or by frost-heaving or the pressure of expanding ice. The easiest way to avoid such problems is to install a drainage system as the wall is constructed.

One effective method is to lay a continuous perforated drain-pipe along the footing of the wall, just as is done with a house foundation footing (Fig. 11-21). The pipeline should be set about 2 inches away from the footing upon a bed of gravel of minimum 2-inch thickness. More gravel is then heaped upon the pipeline, over the footing top and against the bottom of the wall to a depth of about 6 inches. Washed gravel is preferable but not essential, in about a ¾-inch size. The area behind the wall can then be backfilled and compacted to a reasonable degree, up to a point that lies approximately 8 to 12 inches above the proposed finish grade at the face of the wall.

As the wall is built up, weep tubes should be run through masonry at this level and at intervals of about 4 feet. Short lengths of ½-inch PE, PVC, PB or even scraps of CPVC work fine; install them with an appreciable downward slant (toward the face of the wall). The tubes should protrude a half an inch or so from the wall face. At the rear, they should extend about 2 inches into a pocket of ¾-inch gravel of about 1 cubic foot in size, or more. If the wall is a tall one, a second series of weep tubes can be installed about 18 to 24 inches down from the wall top. A good alternative, especially if considerable quantities of moisture can be expected, is to install another drainpipe line, in the same manner as the one at the wall footing, behind the wall about 2 feet below ground level.

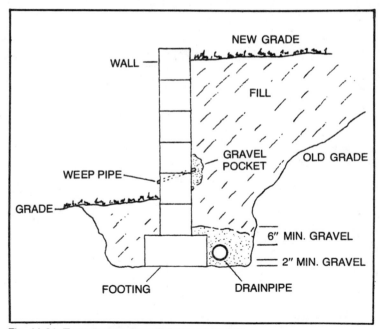

Fig. 11-21. Typical retaining wall drainage system.

As with other drainpipes, those behind retaining walls are best graded at approximately ¼-inch per running foot, especially if large quantities of water will be handled. In most cases it is possible to terminate the pipelines in the open at some convenient spot, so that the moisture can simply drain away. If this is done, the open pipe ends should be closed off with a fairly fine mesh screening to prevent small animals from nesting within the pipes. If open draining is not feasible, the alternative is to pipe the drainwater away into a dry well or an absorption field, either of which can be constructed in the same manner as discussed previously for other types of drainage systems.

Index

405